GLASSES FOR PHOTONICS

This book is an introduction to recent progress in the development and application of glass with special photonics properties. Glass has a number of structural and practical advantages over crystalline materials, including excellent homogeneity, variety of form and size, and the potential for doping with a variety of dopant materials. Glasses with photonic properties have great potential and are expected to play a significant role in the next generation of multimedia systems.

Fundamentals of glass materials are explained in the first chapter and the book then proceeds to a discussion of gradient index glass, laser glasses, non-linear optical glasses and magneto-optical glasses. Beginning with the basic theory, the book discusses actual problems, the performance and applications of glasses. The authors include much useful background material describing developments over the past 20 years in this rapidly moving field.

The book will be of value to graduate students, researchers and professional engineers working in materials science, chemistry and physics with an interest in photonics and glass with special properties.

MASAYUKI YAMANE was born on April 18, 1940 in Kanagawa, Japan. He received his doctor's degree in engineering from Tokyo Institute of Technology in 1968 and joined the academic staff of the Department of Inorganic Materials at Tokyo Institute of Technology in April, 1968, and is now professor. He has worked at Rensselaer Polytechnic Institute, UCLA, the University of Rochester and Université Pierre et Marie Curie. His research is into glass science and technology and sol-gel processing. He has published more than 140 papers on the structure, mechanical properties, optical properties, homogeneity, crystallization, phase separation and sol-gel processing of glasses. He has authored or contributed to 18 books on glass science and engineering. He received the Scientific Award from the Ceramic Association of Japan in 1989.

YOSHIYUKI ASAHARA was born on February 18, 1940 in Fukuoka, Japan. He received his first degree in physics from Tokyo Metropolitan University in 1964 and his doctor's degree in engineering from Kyoto University in 1976. He was with the Materials Research Laboratory of Hoya Corporation from 1964 until 1996 engaged in research and development of new glasses for opto-electronics. His work involved ultrasonic delay line glass, Faraday rotator glass, chalcogenide glasses for optical memory and gradient refractive index fabrication technology. In the past decade his research has been into photonics glass materials, such as nonlinear optical glass and waveguide laser glasses. During this period he was a part-time lecturer at Tokyo Institute of Technology and guest professor at Kyushu University. Dr Asahara has published more than 40 papers on optoelectronics and photonics properties of glasses, and has contributed to 12 books on optoelectronics and photonics. He received the Technical Award from the Ceramic Association of Japan in 1984, the Centennial Memorial Award from the Ceramic Society of Japan in 1991 and the Progress Award from the Laser Society of Japan in 1993.

GLASSES FOR PHOTONICS

MASAYUKI YAMANE
Professor, Tokyo Institute of Technology
AND
YOSHIYUKI ASAHARA
Photonics Technology Advisor

CAMBRIDGE
UNIVERSITY PRESS

CAMBRIDGE UNIVERSITY PRESS
Cambridge, New York, Melbourne, Madrid, Cape Town, Singapore, São Paulo

Cambridge University Press
The Edinburgh Building, Cambridge CB2 2RU, UK

Published in the United States of America by Cambridge University Press, New York

www.cambridge.org
Information on this title: www.cambridge.org/9780521580533

First published 2000
This digitally printed first paperback version 2005

A catalogue record for this publication is available from the British Library

Library of Congress Cataloguing in Publication data
Yamane, Masayuki.
Glasses for photonics/Masayuki Yamane and Yoshiyuki Asahara.
p.   cm.
Includes bibliographical references.
ISBN 0 521 58053 6 (hardback)
1. Glass. 2. Photonics–Materials. I. Asahara, Yoshiyuki, 1940–   . II. Title.
TA450. Y36 2000
621.36–dc21 99-25751 CIP

ISBN-13 978-0-521-58053-3 hardback
ISBN-10 0-521-58053-6 hardback

ISBN-13 978-0-521-01861-6 paperback
ISBN-10 0-521-01861-7 paperback

Contents

Preface

Glass can be doped with rare earth ions, high refractive index ions and micro-crystallites which give it great potential as a photonic medium. The practical advances of these types of glass are expected to have significant roles in multimedia systems in the next generation.

This book is thus an introduction to the theory and recent progress in the technology of glass with special photonic properties, for graduate students, practising engineers and scientists, who wish to supplement their theoretical and practical knowledge of this field with the material science aspects. Hence, this book is intended to be comprehensive enough for an introductory course and be easily readable by practising engineers who are interested in and desire an overview of this field.

Although this book is designed with the purpose of providing a fundamental review of materials with special optical properties, another goal is to provide practical and useful information about developments over the last 10 years in this rapidly changing field. It is impossible, however, to describe all the innovations which have been developed over the last 10 years and omissions are inevitable in a compilation of the size of this book. References to work with respect to the range of glasses examined in this book are given as references to the tables in each chapter. Readers interested in specific data will be able to refer to the original literature. Even so, there remains the possibility of serious omission, for which we beg, in advance, the reader's pardon.

M. Yamane and Y. Asahara

1

Glass properties

Introduction

Glass can be made with excellent homogeneity in a variety of forms and sizes, from small fibers to meter-sized pieces. Furthermore, glass can be doped with rare earth ions and microcrystallites and a wide range of properties can be chosen to meet the needs of various applications. These advantages over crystalline materials are based on the unique structural and thermodynamic features of glass materials. Before discussing the special properties of glass, the fundamentals of glass materials are given in this chapter.

1.1 Features of glass as an industrial material

1.1.1 Structural features

1.1.1.1 Atomic arrangement

A glass is defined in ASTM [1] as 'an inorganic product of fusion which has been cooled to a rigid condition without crystallization'. According to this definition, a glass is a noncrystalline material obtained by a melt-quenching process. Nowadays, noncrystalline materials that can not be distinguished from melt-quenched glasses of the same composition are obtainable by using various techniques such as chemical vapor deposition, sol-gel process, etc. Therefore, most glass scientists regard the term 'glass' as covering 'all noncrystalline solids that show a glass transition' regardless of their preparation method.

The words 'noncrystalline solids' and 'glass transition' suggest that a glass cannot be classified either in the category of crystalline materials such as quartz, sapphire, etc. or in the category of liquid. The atomic arrangement of a glass is different from those of crystalline materials and lacks long-range regularity, as schematically shown in Fig. 1.1 [2]. This is quite close to the

1

Glass properties

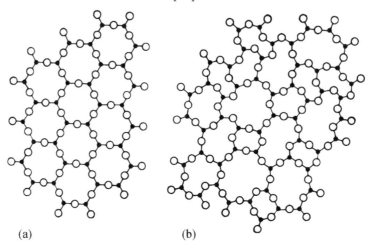

(a)   (b)

Fig. 1.1. Schematic two-dimensional illustration of the atomic arrangement in (a) crystal and (b) glass [2].

atomic arrangement in a liquid. There is neither crystal lattice nor lattice point in the glass structure and therefore, instead of diffraction peaks a halo is seen in the X-ray diffraction patterns of a glass.

Substances which can form noncrystalline solids with the atomic arrangement shown in Fig. 1.1(b) and at an appreciable size are found in oxide, halide, and chalcogenide systems. The three-dimensional random network of strong bonds is developed by the constituent called the 'network former'. Some components called network modifiers can also participate in glass formation by acting to modify the glass properties. These components do not form networks but occupy thermodynamically stable sites as illustrated schematically in Fig. 1.2 or act as a replacement for a part of 'network former'.

Glass formation is possible, in principle, for a system of any composition provided that it contains sufficient of the component called 'network former'. Thus, a wide variety of multi-component glasses can be prepared to attain the desired properties by adjusting the chemical composition at a level below 1%. The lack of regularity of the atomic arrangement over a long range, i.e. the randomness of the structure, is essential to the understanding of the physical and chemical features relevant to those glasses which have the special properties that will be reviewed in this book.

The local environment of the modifying ions in Fig. 1.2 is different from site to site because of the lack in regularity of structure. Since an active ion doped in a glass occupies a similar position to the modifier ions, the absorption and emission spectra from the ion, if any, are broader than those from active ions

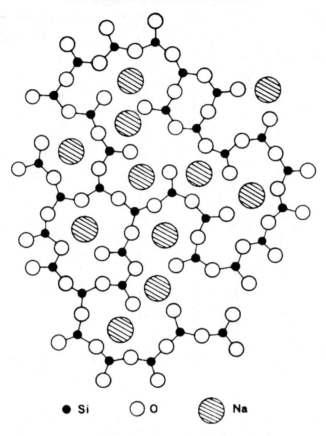

Fig. 1.2. Schematic two-dimensional illustration of the structure of a binary sodium silicate glass [3].

doped in a crystalline material as shown in Fig. 1.3 [4], a feature which is often advantageous in the preparation of a special glass.

The macroscopic properties of a glass such as optical transmission and absorption, refraction of light, thermal expansion, etc. are observed always equally in all directions, provided that the glass is free from stress and strain. That is, a glass is an isotropic material, whereas crystalline materials are generally anisotropic.

1.1.1.2 Chemical composition

Another important feature attributed to the uniqueness of the structure of glass is the flexibility of the chemical composition. Unlike crystalline materials, there is no requirement of stoichiometry among constituents provided that the electrical neutrality over the whole structure of a glass is maintained. Further-

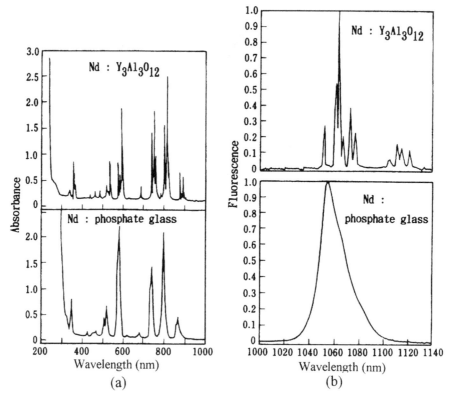

Fig. 1.3. Absorption and emission spectra of Nd ions in YAG crystal and phosphate glass. [Reprinted from M. J. Weber, *Handbook of the Physics and Chemistry of Rare Earth*, Vol. 4, ed. K. A. Gschneider, Jr. and L. Eyring (North-Holland Publishing Company, Amsterdam, 1979) p. 275.]

more, many properties of these multi-component glasses such as molar volume, thermal expansion coefficient, refractive index, etc. can be approximately described using the law of additivity of constituents [5]. These properties reflect the intrinsic nature of the respective constituents in proportion to their contents.

Property, P, of interest can be expressed approximately by Eq. (1.1) using the proportion of respective constituents, f_1, f_2, f_3, ..., f_i, and factors representing their influence, p_1, p_2, p_3, ... p_i, as;

$$P = \Sigma p_i f_i = p_1 f_1 + p_2 f_2 + p_3 f_3 + \ldots + p_i f i. \qquad (1.1)$$

As there is no requirement on the stoichiometric ratio among constituents it is easy to dope active elements such as rare-earth or transition metal elements, semiconductor components, etc. Thus we can obtain a glass which has the unique properties inherent to the doped elements such as special color, emission of fluorescent light, high nonlinear susceptibility, etc.

The other feature that comes from the flexibility of glass composition is the ability to modify the properties of a glass through ion-exchange [6]. There are many stable sites for network modifier ions in a glass structure as shown in Fig. 1.2. For example, alkali ions that are easily thermally activated can move from one stable site to another within a glass. Such a movement of alkali ions within a glass structure enables us to replace alkali ions near the surface of a glass by other ions of the same valence, i.e. Na^+ by K^+; Na^+ by Ag^+; K^+ by $T1^+$, etc. This replacement of alkali ions in the original glass by other ions partially modifies the composition of the glass, and hence its properties. The technique, which will be described later in detail, is particularly important for the modification of optical and mechanical properties.

1.1.2 Thermodynamic features

1.1.2.1 Glass transition

A glass whose atomic arrangement lacks regularity over a long range generally has higher configuration entropy, and hence higher free energy, than a crystalline material of the same composition. In other words, a glass is a thermodynamicaly metastable material which remains un-transformed to its most stable state due to hindrance of the atomic rearrangement during the process of glass formation. The transformation of a glass to a crystal proceeds via nucleation and crystal growth under the driving force of the difference in free energy between crystal and glass, ΔG_v, which increases with the increase of super cooling ΔT_r. Both nucleation rate I_0 and crystal growth rate U are dependent on the viscosity η of a super-cooled liquid as well as on ΔG_v.

Since the viscosity η increases exponentially with decreasing temperature as shown in Fig. 1.4 [7], the influence of viscosity on the hindrance of atomic rearrangement increases with decreasing temperature, whereas the driving force for the transformation into crystal increases with decreasing temperature.

Thus the nucleation rate I_0 and crystal growth rate U are approximately expressed by the following Eq. (1.2) and (1.3) [8, 9]:

$$I_0 = k_1/\eta \exp(-g\alpha^3\beta/T_r(\Delta T_r)^2) \tag{1.2}$$

where k_1 is a constant (typically about 10^{30} dyne cm), $T_r = T/T_m$, $\Delta T_r = (T_m - T)/T_m$, g is a factor related to the shape of the nucleus and equal to $16\pi/3$ for a spherical nucleus, and α and β are dimensionless parameters given by

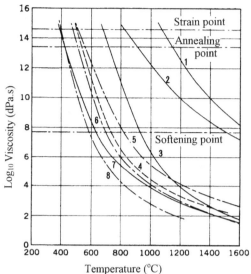

Fig. 1.4. Viscosity–temperature relationship of some commercial glasses. 1: Silicate glass; 2: High (96%) silica glass; 3: Aluminosilicate glass; 4: Soda-lime silicate glass (sheet glass); 5: Borosilicate glass; 6: Soda-lime silicate glass (electric bulb); 7: Lead-alkali silicate glass (electric use); 8: Lead-alkali silicate glass (high lead content). [7] [Reprinted from E. B. Shand, *Glass Engineering Handbook*, (McGraw-Hill Book Company, Inc., 1958) p. 16 copyright (1958) with permission from Corning Inc.]

$$\alpha = (N_A V_C{}^2)^{1/3} \sigma / \Delta H_{fM} \qquad (1.2a)$$

and

$$\beta = \Delta H_{fM} / R T_m$$
$$= \Delta S_{fM} / R \qquad (1.2b)$$

where N_A is Avogadro's number, V_C is the molar volume of the crystal, σ is the crystal–liquid interface energy, ΔH_{fM} is the molar enthalpy of fusion and ΔS_{fM} is the molar entropy of fusion. For typical nonmetals, $\alpha \beta^{1/3}$ is about $0.5 \sim 0.7$.

$$U = f(RT / 3 N_A \pi \lambda^2 \eta)(1 - \exp[\Delta H_{fM} \Delta T / RTTm]) \qquad (1.3)$$

where f is the fraction of the site at which the glass-to-crystal transition can take place on the crystal surface and $\Delta T = T_m - T$, respectively.

In the temperature range near melting point T_m, the influence of $\Delta H_{fM} / T_m$ is predominant and, in the temperature range well below melting point, the influence of viscosity η is very large compared with the influence of $\Delta H_{fM} / T_m$. The schematic illustration of I_0 and U as a function of temperature gives curves which pass through maximum as shown in Fig. 1.5.

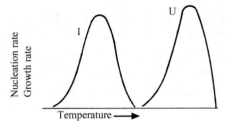

Fig. 1.5.   Schematic illustration of temperature dependence of nucleation rate and crystal growth rate in a super-cooled liquid.

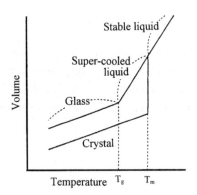

Fig. 1.6.   Volume–temperature relationship of glass-forming liquid, and nonglass-forming liquid.

An important fact to be noted in the figure is that the temperature giving the maximum crystal growth rate (U) is higher than the one giving the maximum nucleation rate (I). When a viscous liquid is cooled from a high temperature, it passes through the temperature region of high crystal growth rate before nucleation takes place. When it passes through the temperature region of high nucleation rate, the crystal growth rate is already very low. Thus, a viscous liquid cooled at a sufficient rate further increases its viscosity to near or above 10^{14} dPa s without being transformed into a crystal. Once the viscosity reaches this value, even the alteration of the local atomic arrangement to equilibrate with temperature is not possible, resulting in the 'frozen-in' structure. Then the material behaves as a rigid and brittle solid, i.e. a glass. Physical and chemical phenomena occurring in a solid glass are no longer so sensitive to temperature change except for those that are enhanced by thermal activation such as the movement of alkali ions. Figure 1.6 shows the temperature dependence of the molar volume of a viscous super-cooled liquid, and uses glass as an example.

The transition from a viscous liquid to a solid glass is called the 'glass transition' and the temperature corresponding to this transition is called the 'glass transition temperature', T_g. The reversible transformation from a glass to a viscous liquid also takes place if a glass is heated to a temperature above T_g.

Since the glass transition occurs as a result of the increase of viscosity and the rate of viscosity increase is dependent on cooling rate, the glass transition temperature, T_g, is not always the same even if the chemical composition is the same, but instead it is usually different, depending on the cooling rate of a liquid. A slow cooling allows enough time for a viscous liquid to alter its local atomic arrangement to attain the minimum free energy at the corresponding temperature, whereas a rapid cooling causes an increase of viscosity that is too quick for the local atomic arrangement to follow and results in a transition into a glass at a higher temperature.

The structure of a rapidly cooled glass is more open than that of a slowly cooled one because the 'freezing-in' of the atomic arrangement occurs at a higher temperature. The properties of a glass are therefore different from glass to glass, depending on the thermal history, even if the chemical composition is the same.

1.1.2.2 Thermal stability and structural relaxation

A glass obtained by cooling a liquid can transform into a crystal if re-heated to a temperature region that is well above T_g, where the nucleation and crystal growth takes place. Figure 1.7 shows an example of the crystallization of a glass detected by a differential thermal analysis (DTA) [10]. The exothermic peak observed in the temperature near 700 °C is attributed to crystallization. The sharpness of the peak and the difference between onset temperature and T_g reflect the thermal stability of a glass. If a glass is thermally unstable, the exothermic peak is sharp and the temperature difference is small.

On the other hand, if the temperature of re-heating is not high and remains near the glass transition temperature, which is well below the temperature giving high crystal growth rate, a glass remains uncrystallized but undergoes some change in atomic arrangement called structural relaxation [11]. The differences in structure and properties between glasses of different thermal history can be eliminated by the structural relaxation brought about by this heat treatment near the glass transition temperature. Figure 1.8 shows an example of such a change attributed to the structural relaxation [12]. It should be noted from this figure that the tuning of some properties of glass is possible by holding the glass at a temperature near the transition temperature.

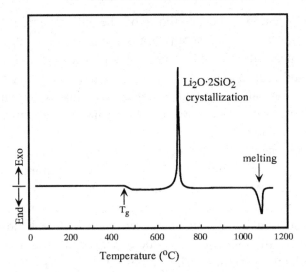

Fig. 1.7.   Differential thermal analysis curve of Li_2O-SiO_2 glass heated at a rate of 10 °C min^{-1} [10].

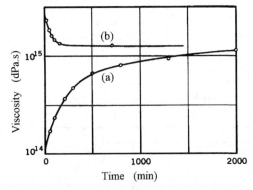

Fig. 1.8.   Viscosity–time curves of a soda-lime silica glass at 486.7 °C [12]. (a) Newly drawn fiber; (b) Fiber stabilized at 477.8 °C for 64 h.

1.1.3 Optical features

1.1.3.1 Transparency

There is neither grain boundary nor interface within a glass structure and so the intrinsic scattering loss of a glass is very small. Therefore, a glass is, in principle, transparent to light in the wavelength region where the intrinsic absorption does not occur, i.e. between two intrinsic absorption edges determined by band-gap energy and the vibration energy of constituents.

Glasses of oxide and fluoride systems that have a wide gap between the conduction and valence bands are generally transparent to light in the visible and near infrared region, whereas those of the chalcogenide system have narrower band gaps and are generally translucent in the visible region but transparent in the near infrared and infrared regions as shown in Fig. 1.9 [13].

Owing to its high transparency to visible light, glass has long been used as the key material for various optical components. Taking advantage of the flexibility of composition, over two hundred kinds of commercial optical glasses have been developed to date.

The primary requisite for an optical glass are high purity and high homogeneity to allow the propagation of a light beam with minimum optical loss. The optical loss inherent to a glass of given composition consists of: (a) intrinsic absorption loss due to the transition of electrons from the valence band to the conduction band which is determined by the band gap energy between two bands; (b) absorption loss due to vibration of molecules; and (c) loss due to Rayleigh scattering. The values of these losses vary with the wavelength of light according to Eqs (1.4), (1.5) and (1.6), respectively [14–16];

$$\alpha(E)_{\mathrm{uv}} = \alpha_0 \exp[A_0(E - E_{\mathrm{g}})/kT] \tag{1.4}$$

where E is phonon energy, $\alpha(E)_{\mathrm{uv}}$ is the absorption coefficient at photon energy E, due to intrinsic absorption loss in UV region, E_{g} is band gap energy between the valence band and conduction band, k is Boltzmann constant, and α_0 and A_0 are constants [14].

$$\alpha(E)_{\mathrm{IR}} = \alpha_2 \exp(-E/E_2) \tag{1.5}$$

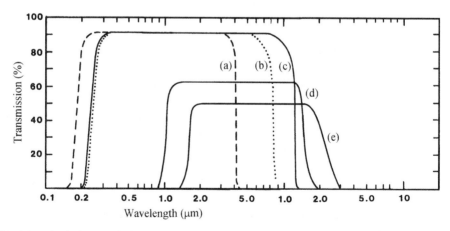

Fig. 1.9. Optical transmission curves of representative oxide, fluoride and chalcogenide glasses. (a) Silica glass, (b) $57HfF_4-36BaF_2-3LaF_3-4AlF_3$ (mol%) glass; (c) $19BaF_2-27ZnF_2-27LuF_3-27ThF_4$ (mol%) glass; (d) As_2Se_3 glass; (e) $10Ge-50As-40Te$ (atom%) glass. [Reprinted from M. G. Drexage, *Treatise on Materials Science and Technology*, Vol. 26, Glass IV (Academic Press, Inc., 1985) p. 151, copyright (1985) with permission from Academic Press, Inc.]

where, $\alpha(E)_{IR}$ is the absorption coefficient at photon energy E due to vibration of molecules, and α_2 and E_2 are constants [15].

$$I_{RS}(\lambda) = (8\pi^3/3\lambda^4)[(\partial n^2/\partial \rho)^2(\delta \rho)^2 + \Sigma(\partial n^2/\partial C_i)^2(\delta C_i)^2]\delta V \qquad (1.6a)$$

where

$$(\delta \rho)^2/\rho^2 = (\beta_c/\delta V)kT_F \qquad (1.6b)$$

$$(\delta C_i)^2 = kT_F/(N_i \partial \mu_c/\partial C_i) \qquad (1.6c)$$

and $\delta \rho$ is density fluctuation, δC_i is compositional fluctuation of the ith component, β_c is isothermal compressibility, T_F is the fictive temperature, k is Boltzmann constant, N_i is number of molecules of ith component contained in the unit volume of glass, μ_c is the chemical potential, and n is the refractive index of glass respectively [16].

Figure 1.10 depicts the change of respective losses with wavelength of light in the case of silica glass [17]. It is well known that the theoretical minimum loss is determined by the absorption loss due to vibration of molecules and Rayleigh scattering, and is about 0.125 dB km^{-1} in the vicinity of 1.55 μm.

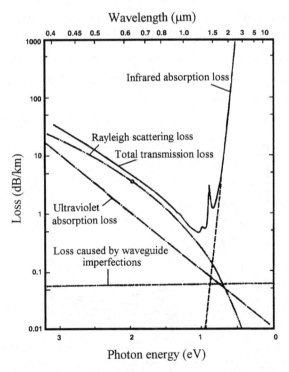

Fig. 1.10. Changes of intrinsic optical losses of P_2O_5-doped SiO_2 glass with wavelength of light. [Reprinted from T. Miya, T. Terunuma, T. Hosaka and T. Miyashita, *Electron. Lett.* **15** (1979) 106, copyright (1979) with permission from the Institute of Electronic Engineers.]

It should be noted that the wavelength giving the minimum intrinsic loss is different from glass to glass depending on the type and amount of the constituents. The loss value of a practical glass is the sum of this intrinsic loss and the extrinsic loss attributed to the absorption by impure atoms and the scattering loss due to compositional heterogeneity, etc. Major impurities that cause absorption loss are transition metal ions and water. Absorption by transition metal ions is attributed to the transition of electrons between the d-orbital of two different energy levels determined by the influence of the ligand of the ion. The absorption spectra assigned to various transition metals in a silica glass are shown in Fig. 1.11 as examples [18].

The absorption loss due to water in a glass is attributed to the stretching vibration of the OH group and its coupling with the vibration of the metal–oxygen bond. The wavelength and intensity of such absorption in silica glass are shown in Table 1.1 as an example [19].

1.1.3.2 Linear and non-linear refractive index, and dispersion

Refractive index is another important property to be considered with respect to the optical features of glass. It is correlated with the electric dipole moment induced by the electromagnetic interaction of constituent atoms and molecules with light, and is expressed by Eq. (1.7) which is known as the Lorentz–Lorenz equation as [20]:

$$(n^2 - 1)V/(n^2 + 2) = (4\pi/3)N_A \alpha$$
$$= R \qquad (1.7)$$

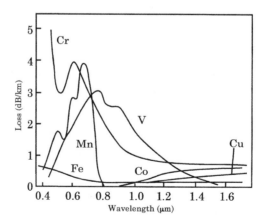

Fig. 1.11. Absorption spectra of various transition metal ions in a silica glass. [Reprinted from P. C. Shultz, *J. Am. Ceram. Soc.* **57** (1974) 309, with permission of The American Ceramic Society, Post Office Box 6136, Westerville, Ohio 43086-6136, copyright (1974) by The American Ceramic Society. All rights reserved.]

Table 1.1. *Absorptions assigned to vibration of OH and their coupling with vibration of metal-oxygen bonds [Reprinted from P. Kaiser, A. R. Tynes, H. W. Astle, A. D. Pearson, W. G. French, R. E. Jeagar et al., J. Opt. Soc. Am. **63** (1973) 1141, copyright (1973) with permission from the Optical Society of America]*

Wavelength λ (μm)	Vibration mode	Absorption loss (dB/km/ppm of OH)
2.73	v_3	8300
2.22	$v_1 + v_3$	220
1.93	$2v + v_3$	8.6
1.39	$2v_3$	54
1.24	$v_1 + 2v_3$	2.3
1.13	$2v_1 + 2v_3$	0.92
0.945	$3v_3$	0.83
0.88	$v_1 + 3v_3$	7.5×10^{-2}
0.82	$2v_1 + 3v_3$	3.3×10^{-3}
0.72	$4v_3$	5.8×10^{-2}
0.68	$v_1 + 4v_3$	3.3×10^{-3}
0.64	$2v_1 + 4v_3$	8.3×10^{-4}
0.60	$5v_3$	5.0×10^{-3}

v_3, Stretching vibration of Si-OH;
v_1, Stretching vibration of Si-O.

where V is molar volume, n is refractive index, α is polarizability, N_A is Avogadro's number, and R is molar refraction.

When the number of vibrating electrons in the presence of an electric field E, which changes periodically with frequency ω, is large, the polarizability α is expressed as:

$$\alpha = (e^2/4\pi^2 m)\Sigma[f_i/(\omega_i^2 - \omega^2)] \tag{1.7a}$$

where e is charge of an electron, m is mass of electron, ω_i is the frequency of vibration of an electron in the ith ion, and f_i is the oscillator strength of the intrinsic absorption of the ith oxygen ion, respectively. The refractive index of a glass is then written as:

$$(n^2 - 1)/(n^2 + 2) = N_A(e^2/3\pi m)\Sigma[f_i/(\omega_i^2 - \omega^2)]. \tag{1.7b}$$

The values of ω_i of a material which are non-absorbing in the visible region of the spectrum lie within the range of the shorter wavelengths and $\omega_i > \omega$. The refractive index n of a glass given by this equation is in the range 1.45 to 2.00 depending on the composition of the glass and decreases gradually with the increase of wavelength of light in the visible region, as shown in Fig. 1.12 [21]. This dependence of refractive index on wavelength of light is called dispersion and is also an important parameter used to describe the optical properties of a glass.

The dispersion of a glass is usually represented either by 'average dispersion' $n_F - n_C$ or by the Abbe number ν_d given by Eq. (1.8).

$$\nu_d = (n_d - 1)/(n_F - n_C) \tag{1.8}$$

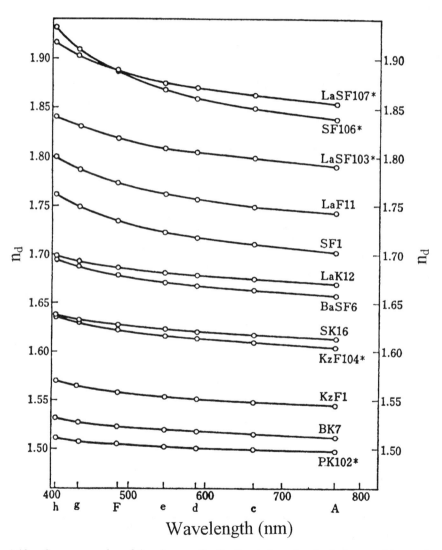

Fig. 1.12. Some examples of the change of refractive index of optical glasses with wavelength of light [21].

where n_d, n_F, n_C are the refractive indices for the d-line of He (wavelength: 587.56 nm), the F-line of H (wavelength 486.13 nm) and the C-line of H (wavelength 656.27 nm), respectively.

The relations described above, or the refractive index of ordinary optical glasses used for lenses, prisms, etc., apply within the range where the linear relationship, given by Eq. (1.9), holds between the polarization P and electric field E.

$$P = \chi^{(1)} E \tag{1.9}$$

where the proportionality coefficient $\chi^{(1)}$ is called linear susceptibility. Under high electric fields such as those caused by the irradiation of strong coherent light such as laser beams, however, the induced polarization does not change linearly with applied field but increases nonlinearly as expressed by Eq. (1.10) [22]

$$P = \chi^{(1)} E + \chi^{(2)} E^2 + \chi^{(3)} E^3 + \cdots \tag{1.10}$$

where, the coefficients $\chi^{(2)}$ and $\chi^{(3)}$ are called second- and third-order nonlinear susceptibilities, respectively.

In an isotropic material like glass, the second-order nonlinear susceptibility $\chi^{(2)}$ becomes zero and the nonlinearity at orders higher than four is negligibly small. The Eq. (1.10) is rewritten as Eq. (1.10′) for an ordinary glass

$$P = \chi^{(1)} E + \chi^{(3)} E^3. \tag{1.10′}$$

Consequently, the refractive index of a glass n also varies with the intensity I of light as:

$$n = n_0 + (12\pi\chi^{(3)}/n_0)E^2 = n_0 + n_2'' I \tag{1.11}$$

where n_0 is the linear refractive index when the relation in Eq. (1.9) holds, and n_2'' is the nonlinear refractive index. Application of this nonlinear effect is one of the key technologies in the field of optics and optoelectronics and will be reviewed in Chapter 4.

1.2 Classification of glasses by preparation method

1.2.1 Glass by melt-quenching technique

Before the development of the chemical vapor deposition and sol-gel processes, the melt-quenching technique was the only method by which bulk glasses of acceptable size for practical application could be obtained. Even today, glasses produced by the melt-quenching technique make up more than 99% of practical glasses in both volume and number of types. The process, which is based on the fusion of crystalline raw materials into a viscous liquid followed by forming it into a shape and quenching to a glass [23], is distinguished from other methods of glass preparation in many aspects including the available systems, size and shape of the products, number of components, etc.

A batch mixture prepared by mixing predetermined amounts of pulverized crystalline raw materials to attain the desired property is placed in a tank furnace made of firebrick or into a crucible made of clay, platinum, etc. to be fused into a liquid at high temperature. A tank furnace is designed to allow the continuous mass production at a daily pull of $40 \sim 600$ tons of glass by charging the batch mixture from one end and taking the molten glass out of the other end of the furnace to feed to a forming machine.

When glass which has a special property is required in a limited amount, the melting of a batch mixture is made using a crucible with an externally supplied heat source such as the combustion of fossil fuel or electricity using an induction coil, and then supplied to a forming machine. Depending on the application, the melt is sometimes cooled very slowly to a rigid condition within a clay crucible so that only the highly homogeneous part is used. Generally the molten glass is kept at the temperature which corresponds to a viscosity of $10 \sim 10^2$ dPa s, i.e. $1400 \sim 1500$ °C for most commercial glasses as shown in Fig. 1.4, for more than 10 h in order to remove tiny bubbles and to enhance the homogenization of the melt through convection and the interdiffusion of the constituent atoms.

The formation of a melt into a desired shape is carried out at a temperature corresponding to the viscosity of $10^3 \sim 10^4$ dPa s, i.e. $900 \sim 1100$ °C for most commercial glasses, by applying various forming methods such as casting into a mold, blowing, up-drawing, down-drawing, pressing, rolling out, floating, and various combinations, and in this way the high flexibility in the geometry of products, which is the most distinctive feature of the melt-quenching technique, is possible. It is well known that a continuous glass ribbon of about 5 m wide and $0.5 \sim 20$ mm in thickness is routinely produced in the sheet glass industry throughout the world; a continuous glass fiber of several microns in diameter is drawn through a platinum nozzle; and an optical glass for telescope lens of the diameter exceeding 1 m is easily obtainable.

Most of the formed glasses are usually annealed in a temperature range that is slightly higher than the glass transition temperature (shown as the annealing point in Fig. 1.4) to remove any thermal stress which developed during the forming and subsequent cooling due to the low thermal conductivity of a glass.

The additional feature of the melt-quenching method, namely the high flexibility of the geometry of a glass and particularly the advantage in obtaining materials of large size in comparison with a single crystal or polycrystalline ceramics, is as important as the structural and thermodynamic features when

Table 1.2. *Chemical composition of some industrial glasses manufactured by melt-quenching technique (wt%) [24]*

Glass	SiO_2	Al_2O_3	B_2O_3	CaO	MgO	Na_2O	K_2O	PbO	Other
Sheet glass by float	72.6	1.8	–	7.9	3.8	12.2		–	0.7
Container glass	72.8	1.6	–	11.3		13.8		–	0.5
Crystal glass	56.9	–	–	–	–	2.8	13.6	26.0	0.3
Optical glass SF6	26.9	–	–	–	–	0.5	1.0	71.3	0.3
Optical glass BK7	66	–	12.4	–	–	8	12	–	1.6
Laboratory ware	80	2.25	13	–	–	3.5	1.15	–	0.9
Electric glass (stem)	57	1	–	–	–	3.5	9	29	0.5
TV panel glass	63.8	1.3	9.4	2.9 (SrO)	2.2 (BaO)	7.2	8.8	2.8	1.6
TV funnel	51	4	–	8[a]	–	6	7.5	23	0.5
TV neck	47	3.3	–	3.4[a]	–	3	10	33	0.3

[a]RO (= CaO + BaO + SrO).

considering the preparation of glasses with special properties. Good examples of the products based on this feature are the giant pulse laser, Faraday rotator, etc., whose total performance is dependent on the size of material.

The other advantage of the melt-quenching technique over chemical vapor deposition or the sol-gel process is the large flexibility of composition. Since simple quenching of a melt does not require stoichiometry among constituents, the preparation of glasses with a wide variety of compositions, consisting of sometimes up to ten kinds of constituents at various ratios from a few to several tens of percent, is possible. The doping or co-doping of active ions such as rare-earth or transition metal at a level of a few percent or less is also made relatively easily, which is quite important for the production of glasses with special properties. As shown in Table 1.2, most of the important industrial glasses are based on this advantage of flexibility of composition. Examples of the glasses are color filters, and various glass lasers, etc.

The important fact that should be stressed about the melt-quenching technique is that most of the above-mentioned features are true not only for the familiar silicate, borate, or phosphate systems but also for the many exotic glasses of the oxide system as well as non-oxide glasses such as those of the fluoride and metal alloy systems. It should, however, be noted that there are some disadvantages in the melt-quenching technique compared with other processes. The melt-quenching technique which uses pulverized crystalline raw materials has disadvantages for the preparation of glasses of ultra-high purity such as those used for optical communication in comparison with the chemical vapor deposition and sol-gel processes that use liquid raw materials. The

Table 1.3. *Examples of metal halides and their boiling points employed in glass formation by chemical vapor deposition process [27]*

Formula	Melting point (°C)	Boiling point (°C)	Density (g cm^{-3})
SiH_3Cl	-118	-30.4	1.15
SiH_2Cl_2	-122	8.3	1.42
$SiHCl_3$	-127	31.8	1.35
$SiCl_4$	-70	57.6	1.48
$GeCl_4$	-49.5	84.6	1.84
PCl_3	-93.6	74.7	1.57
$POCl_3$	1.3	105.1	–
BBr_3	-46.0	96.0	–
$AlBr_3$	97.5	263.3	–
$TiCl_4$	-25	136.4	1.726

difficulty of maintaining high purity is also attributed to possible contamination by the crucible or furnace materials which often react with the glass melt under high temperature. In order to minimize the contamination, crucibles made of noble metal such as platinum and its alloys are often used for the preparation of special glasses.

Another disadvantage of the melt-quenching technique that should be noted is that it is extremely difficult to prepare glasses containing a large amount of refractory material represented by SiO_2, TiO_2, Al_2O_3, ZrO_2 etc. by this technique because of the requirement for an extremely high temperature for melting. The production of the binary or ternary glasses containing these components, as well as silica glass of ultra-high purity, are therefore made by either chemical vapor deposition or the sol-gel process.

1.2.2 Glass by chemical vapor deposition

The preparation of a bulk glass by the chemical vapor deposition (CVD) method was developed in the early 1940s [25]. The process is based on the thermally activated homogeneous oxidation or hydrolysis of the initial metal halide vapor (or mixture of metal halides) to form particulate glass material 'soot', followed by viscous sintering of the soot into solid inclusion-free glass bodies [26]. Examples of the metal halides employed in this process are shown in Table 1.3 along with their melting and boiling points [27].

The oxidation or hydrolysis reaction is usually activated by either oxygen plasma or an oxy-hydrogen flame. In the case of the formation of silica glass

Table 1.4. *Melting point and boiling point of chlorides of transition metals, rare-earth elements and some alkali and alkaline earth metals [27]*

Formula	Melting point (°C)	Boiling point (°C)
$FeCl_3$	300	317
$CuCl_2$	422	1366
$CrCl_3$	1150	–
$NiCl_3$	1001	–
$MnCl_3$	650	1190
$CoCl_3$	500	1049
$EuCl_2$	727	>2000
$ErCl_3 6H_2O$	774	1500
$NdCl_3$	784	1600
$SmCl_3$	678	–
$LiCl$	605	1325–1360
$NaCl$	801	1413
$CaCl_2$	782	>1600
$BaCl_2$	963	1560

from $SiCl_4$, the reaction is schematically shown by Eq. (1.12) and (1.13), respectively

$$SiCl_4 + O_2 = SiO_2 + Cl_2 \qquad (1.12)$$
$$SiCl_4 + H_2O = SiO_2 + HCl. \qquad (1.13)$$

The typical reaction temperature in the former case is about 2000 °C, whereas the sintering temperature of the soot into a glass is about 1800 °C [28]. The most distinctive difference between the reaction using oxygen plasma and that using an oxy-hydrogen flame is the level of OH contained in the eventual glass. The hydroxyl level in the silica glass produced by using oxygen plasma is as low as 1 ppm OH, whereas the typical hydroxyl level in a silica glass by the latter method is 1000 ppm OH.

The feature of this method is that the initial metal halides are liquid at or in the vicinity of room temperature and their boiling points are very low compared with the boiling points of the halides of alkali, alkaline earth, transition metals, or rare-earth elements, examples of which are shown in Table 1.4 [27]. This means that the purification of raw materials is easily made by repeating the distillation at a temperature well below the melting points of the halides in this table.

Thus, the process has many advantages over the melt-quenching technique for preparing glasses of ultra high purity, particularly the preparation of glasses that are free from transition metal elements. In addition, it is possible to prepare glasses whose chemical composition can be varied continuously or step-wise by changing the mixing ratio of the raw materials. The method, however, is not suitable for the preparation of glasses containing alkali and alkaline earth elements or for glasses doped with rare-earth elements.

The types of special glasses that are produced by using the advantage of this process include high-purity silica glass used for various optical and opto-electronic devices [26], TiO_2-SiO_2 glass of ultra-low thermal expansion used for the mirror blanks for telescopes [29], the preforms of optical fiber for telecommunications [30] etc.

An optical fiber produced by the chemical vapor deposition process is 125 µm in diameter and has a core-clad structure as shown in Fig. 1.13. The refractive index of the core is slightly ($0.1 \sim 0.3\%$ and $0.8 \sim 2\%$ depending on the type of the fiber) higher than that of the cladding. The optical signal enters the fiber and is propagated within the core along the fiber axis. The fiber is classified into three types depending on the refractive index profile of the core-clad structure which determines the mode of propagation of the optical signal.

The single mode fiber whose core diameter is $8 \sim 10$ µm allows the

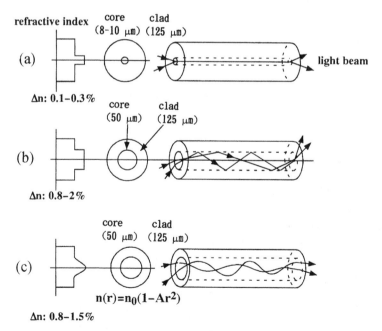

Fig. 1.13.   Structure of optical fiber for telecommunications. (a) Single-mode fiber. (b) Multi-mode fiber of step index type. (c) Multi-mode fiber of graded index type.

propagation of the signal of only one mode. Its transmission capacity is much higher than those of other two types and it is now used for the majority of the fiber for telecommunication. The multi-mode fiber having a core diameter of 50 μm allows the propagation of the signal in multiple modes. There are two types of multi-mode fiber depending on the profile of the refractive index in the core glass. The graded type is designed to compensate for the difference in the propagation path length among various modes so that the distortion of the signal is minimized.

The fabrication of the preform for optical fiber by chemical vapor deposition has been carried out to date by three different methods, the OVD (Outside Vapor-phase Deposition) [31], MCVD (Modified Chemical Vapor Deposition) [32] and VAD (Vapor-phase Axial Deposition) [33] processes. Schematic illustrations of these methods are shown in Fig. 1.14.

In the OVD process, a soot of desired glass composition is deposited on a rotating and traversing ceramic target rod by passing the vapor stream through a methane–oxygen flame directed at the target rod. The glass soot sticks to the rod in a partially sintered state and a cylindrical porous glass preform is built up. The core-clad structure is formed by supplying the metal halide vapor to give a high refractive index such as $GeCl_4$ or $POCl_3$ along with $SiCl_4$ first until the desired thickness is obtained, and then switching into the stream of $SiCl_4$ alone. When the soot deposition is completed, the porous glass preform is

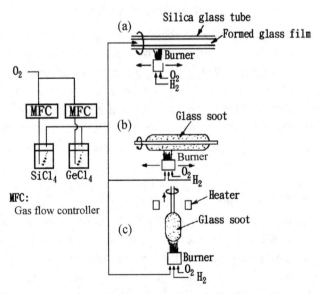

Fig. 1.14. Fabrication methods of preform of optical fiber. (a) MCVD process. (b) OVD process. (c) VAD process.

slipped off the target rod to be sintered into a solid glass blank by passing through a furnace hot zone at a temperature of about 1600 °C. The central hole which still remains at this stage disappears when it is drawn into fiber at a temperature of about 1800 ∼ 1900 °C.

In the MCVD process, the mixture of $SiCl_4$ and $GeCl_4$ (or $POCl_3$) is supplied, together with oxygen, to a rotating silica glass tube on a lathe to form the glass soot and sinter it into a glass layer by heating with a traversing burner. The silica glass tubing that has a glass layer of higher index in its inside is then collapsed by zone sintering into a solid glass preform consisted of a core of SiO_2-GeO_2 (or $SiO_2-P_2O_5$) glass and a clad of the original silica glass tubing.

In the VAD process, a soot having the desired glass composition is deposited on the bottom end of a rotating silica glass target rod by passing the vapor stream through an oxy-hydrogen flame directed at the target rod. The process involves the simultaneous flame deposition of both core and cladding glass soot as the preform grows. When the soot deposition is completed, the porous glass preform is consolidated to solid glass by zone sintering. In principle, the process should allow the continuous making of blanks.

The optical loss of a fiber drawn from a preformed glass rod prepared at a temperature of about 2000 °C is less than 0.5 dB km^{-1} at the wavelength of 1.3 μm and the fiber can transmit the optical signal as far as 40 km without a repeater.

1.2.3 Glass by sol-gel process

Glass preparation by sol-gel process has been extensively studied since H. Dislich [34] reported the formation of borosilicate glass by hot-pressing granules of dehydrated gel in 1971. The sol-gel process begins with formation of a sol, consisting of colloids dispersed in a liquid medium. A sol turns into a porous gel by the coagulation of these colloids while standing in a mold or on a substrate which is used as a coating film, etc. The gel thus obtained is dried and sintered into a pore-free dense glass or glass film, etc. at a temperature slightly above the glass transition temperature of the eventual glass [35].

There are various ways of preparing a sol which depend on the material to be made. In the case of the formation of bulk glass, the most widely used method is the hydrolysis and polycondensation of silicon alkoxide or its mixture with the alkoxides of Ti, Al, Zr, Ge, etc. An alternative way is to disperse fine particles of sub-micron size obtained by flame hydrolysis of $SiCl_4$ in chloroform using n-propanol as dispersing agent. Representative metal alkoxides employed in the glass preparation by hydrolysis and polycondensation are shown in Table 1.5 along with their boiling points [36].

Table 1.5. *Representative metal alkoxides employed in glass preparation [36]*

Alkoxide	Formula	Boiling Point (°C/mmHg)
Boron *n*-butoxide	$B(OBu^n)_3$	128/760
Aluminium *n*-propoxide	$Al(OPr^n)_3$	205/1.0, 245/10.0
Aluminium *n*-butoxide	$Al(OBu^n)_3$	242/0.7, 284/10.0
Aluminium *tert*-butoxide	$Al(OBu^t)_3$	151/1.3, 185/10.0
Silicon ethoxide (TEOS)	$Si(OEt)_4$	168/760
Silicon methoxide (TMOS)	$Si(OMe)_4$	153/760
Germanium ethoxide	$Ge(OEt)_4$	86/12.0
Titanium ethoxide	$Ti(OEt)_4$	103/0.1
Titanium *tert*-butoxide	$Ti(OBu^t)_4$	93.8/5.0
Zirconium ethoxide	$Zr(OEt)_4$	190/0.1

Me, CH_3; Et, C_2H_5; Pr, C_3H_7; *n*-Bu, C_4H_9; *tert*-Bu, $(CH_3)_3C$.

The hydrolysis and condensation reactions of these metal alkoxides are schematically expressed by the Eq. (1.14) and (1.15).

$$Me(OR)_n + nH_2O = Me(OH)_n + nROH \qquad (1.14)$$

$$mMe(OH)_n = mMeO_{n/2} + (mn/2)H_2O \qquad (1.15)$$

where Me is Si, Al, Ti, Ge, B, Zr, etc., R is CH_3, C_2H_5, C_3H_7, C_4H_9, etc. The basic procedure to form, for example, a silica glass rod using silicon methoxide, TMOS, as the starting material is schematically illustrated in Fig. 1.15 [37].

Initially 1 mol of TMOS is dissolved in about 5 mol of methanol contained in a flask under stirring and then 4 mol of water is added to the solution under vigorous stirring. The role of the methanol is to form a homogeneous solution

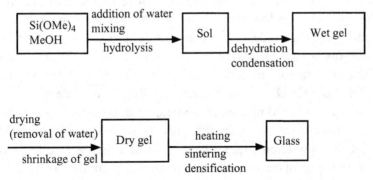

Fig. 1.15.   Schematic illustration of sol-gel process for glass preparation.

Table 1.6. *Some properties of silica glass prepared*
from silicon methoxide [37]

Property	Sol-gel derived silica glass	Commercial silica glass
Density $(g\,cm^{-3})$	2.20	2.21
Refractive index (n_d)	1.457	1.4584
Vickers hardness	660–700	680–700
Thermal expansion $(°C^{-1})$	5.5×10^{-7}	5.5×10^{-7}
Young's modulus (K bar)	700–720	730
Bulk modulus (K bar)	310	314

as the mutual solvent of two immiscible liquids, TMOS and water. The clear silicate solution is cast in a cylindrical plastic mold and left in an oven at a temperature that is below the boiling point of methanol, for example at about 65 °C, with a tightly sealed cover to enhance gel formation through the hydrolysis and condensation reaction expressed by Eqs (1.14) and (1.15), followed by aging at the same temperature until it separates from the container wall by the syneresis of a small amount of liquid.

The gel is then dried by slowly evaporating the liquid through, for example, pin holes formed on the cover until no more shrinkage occurs. Slow drying is necessary to avoid fracture of the gel due to a capillary force induced by the evaporation of the liquid.

The dried silica gel is, then, subjected to heat treatment under vacuum at a temperture about 1000 °C for densification into a pore-free glass. The treatment under vacuum is to enhance the removal of water that was generated by the dehydration of silanols covering the surface of the silica particles. Some properties of silica glass prepared by this method are compared with those of commercial silica glass in Table 1.6 [37].

It should be noted that the highest temperature required to obtain a silica glass is 1000 °C, which is lower by about 1000 °C than the temperature required for the melt-quenching process. Thus, the sol-gel process has great advantages over the melt-quenching technique for the preparation of glasses containing a large amount of refractory materials such as Al_2O_3, TiO_2, ZrO_2, GeO_2, etc. In addition, unlike the CVD process it is also possible to prepare glasses containing alkali, alkali-earth, and rare-earth elements, etc. by using aqueous solutions of their salts instead of water in the process shown in Fig. 1.15. Another advantage of this low-temperature process is that it is possible to incorporate thermally unstable compounds such as nonoxide semiconductors into a glass to develop a glass with special properties such as high optical nonlinearity.

Since most alkoxide precursors are in a liquid state and are easily purified by repeating distillation, and since there is no possibility of contamination from a container because this is a crucibleless process, it is also possible to obtain glasses of very high purity such as those used for fabrication of optical fiber.

Preparation of special glasses which are difficult to produce by the melt-quenching or chemical vapor deposition process is also possible by using special treatment such as partial leaching or partial replacement of constituents of a gel through micropores. A good example of a product based on this advantage is a gradient index (GRIN) rod of large diameter [38] which will be reviewed in Chapter 2.

There are of course some drawbacks or disadvantages compared with the two other processes discussed above. The biggest drawback of the sol-gel process is the large amount of shrinkage of a wet gel upon drying which often leads to a fracture. This excessive shrinkage is caused by a capillary force which is induced by the evaporation of liquid in micropores and is expressed by Eq. (1.16).

$$P = 2\gamma \cos \theta / D \qquad (1.16)$$

where P is capillary force, γ is surface tension of liquid, θ is the contact angle of liquid, and D is diameter of micropores.

A tremendous amount of effort has been made during the past two decades by various researchers to hinder gel fracture during drying. These include: (a) the enhancement of hydrolysis with acidic water of pH about 2, and the subsequent addition of ammonia water of high pH to enhance the condensation reaction in order to obtain a gel with large pores [39]; (b) the introduction of an organic solvent called a 'drying control chemical additive, (DCCA)' such as formamide and di-methylformamide, having a lower surface tension and a higher boiling point than water, together with alcohols [40]; (c) drying the gel under hypercritical conditions where the liquid–vapor interface, and therefore the capillary force, diminishes [41]; (d) the incorporation of separately pre-pared silica glass particles of sub-micron size in the alkoxide–water–alcohol solution to form a gel of bimodal pore-size distribution [42]. Using these effects, a silica glass rod of about 30 mm in diameter and 150 mm in length is now obtainable [43].

The other big problems which had to be overcome were the preferential precipitation of a particular oxide during sol formation for multicomponent glasses made from alkoxide precursors, and the heterogeneous precipitation of metal salts on the surface during the drying of a gel which was made using aqueous solutions of metal salts as the source of alkali, alkali-earth oxides, etc.

Preferential precipitation of a particular oxide during sol formation is due to

Table 1.7. *Electronegativity 'χ', partial charge 'δM',
ionic radius 'r', and coordination number 'n' of some
tetravalent metals (z = 4) [Reprinted from J. Livage, F.
Babonneau and C. Sanchez, Sol-Gel Optics, Processing and Applications, ed. L. C. Klein (1994) 39,
copyright (1994) with permission from Kluwer
Academic Publishers]*

Alkoxide	χ	δM	r(A)	n
$Si(OPr^i)_4$	1.74	+0.32	0.40	4
$Ti(OPr^i)_4$	1.32	+0.60	0.64	6
$Zr(OPr^i)_4$	1.29	+0.64	0.87	7
$Ce(OPr^i)_4$	1.17	+0.75	1.02	8

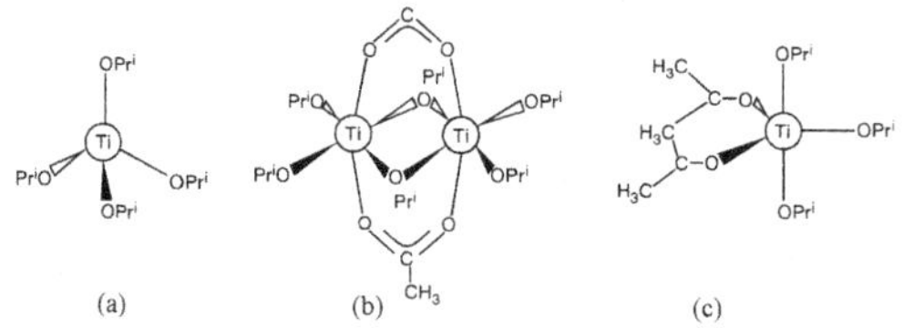

(a)   (b)   (c)

Fig. 1.16. Chemical modification of transition metal alkoxides. (a) $Ti(OPr^i)_4$. (b) $[Ti(OPr^i)_3(OAc)]_2$. (c) $[Ti(OPr^i)_3(acac)]$. [Reprinted from J. Livage, F. Babonneau and C. Sanchez, *Sol-Gel Optics, Processing and Applications*, ed. L. C. Klein (1994) 39, copyright (1994) with permission from Kluwer Academic Publishers].

the difference in the reactivity of metal alkoxides toward hydrolysis, which, in turn, depends mainly on the positive charge of the metal atoms $δM$ and its ability to increase its coordination number 'n'. As a general rule the electro-negativity of a metal atom decreases and the ability to increase its coordination number 'n' increases when going down the periodic table as shown in Table 1.7 [44]. Accordingly, the chemical reactivity of the corresponding alkoxides increases as their size increases.

Silicon retains fourfold coordination ($n = 4$) in the molecular precursor as well as in the oxide. The electronegativity of Si is rather high and its positive charge is small. Silicon alkoxides are therefore not very sensitive to hydrolysis whereas alkoxides of Ti, Zr, Ce are very sensitive to moisture due to the low electronegativity of their atoms and the increase of coordination number by the hydrolysis. One of the ways to control the reactivity of the alkoxides of Ti, Zr, Ce, etc. is to chemically modify the coordination state of the alkoxides with a chelating agent such as acetylacetone as shown in Fig. 1.16.

The other way to hinder the preferential precipitation of a particular oxide is to partially hydrolyze the less sensitive silicon alkoxide first and then add the sensitive alkoxide to facilitate the condensation reaction with the partially hydrolyzed silicon alkoxide [45].

The heterogeneous precipitation of metal salts on the surface during drying is due to the migration of constituent ions and water towards the gel surface as the dehydration advances. An effective method to hinder this crystallization is to replace the liquid in the micropores of a gel with an organic solvent, which is miscible with water but does not dissolve the metal salts, so that the salt is precipitated and fixed as micro-crystallites on the micropore wall at the front of the penetrating organic solvent [46]. This treatment is particularly important in the preparation of alkali-containing multi-component glasses. In addition to the bulk glasses discussed above, a wide variety of thin films of thickness $0.1 \sim 0.5$ μm have been successfully prepared by dip- or spin-coating of an alkoxy-derived sol on a glass substrate to modify glass properties [47].

1.3 Important glass systems

1.3.1 Oxide glasses

1.3.1.1 Silica glass

Among various oxides used in the industrial materials, SiO_2, GeO_2, B_2O_3, and P_2O_5 are known to be good network formers which can develop the three-dimensional random network shown in Fig. 1.16 and can form a glass by itself. Only silica glass has been practically used to date as a single component glass.

A silica glass shows a superior refractory nature, thermal shock resistance [48], optical transparency in both the UV and visible regions [49], and high chemical durability. Due to these properties, silica glass has been widely used in the semiconductor industry as a crucible for the formation of silicon, a mask blank for photolithography, as well as various optical elements in optic and opto-electronic devices [26].

The drawback of this glass as an industrial material, if any, is high cost due to its very high viscosity which requires a temperature above 2000 °C for continuous mass production if the melt-quenching technique is applied. Commercial silica glass is supplied in the form of tubing, plate, or rod and re-formed by heating to a temperature around 1800 °C, into the desired shape depending on the application. Silica glass is classified into five types according to the production method and the level of impurities [48].

Type-I silica glass is produced by electrically fusing natural quartz in a vacuum or in an inert gas atmosphere. It contains Al_2O_3 and Na_2O as impurities at the level of 30 $\sim$ 200 ppm and 4 $\sim$ 10 ppm, respectively. Type-II silica is produced by fusing pulverized natural quartz using an oxyhydrogen flame. The content of Al_2O_3 and Na_2O in this glass is lower than those in Type-I silica, but the glass contains hydroxyl group impurities from the melting atmosphere, at the level of 250 $\sim$ 400 ppm.

Type-III silica is synthesized by the flame hydrolysis of $SiCl_4$ according to the reaction shown by the Eq. (1.13) in the previous section. The glass has very high homogeneity and is almost free of Al_2O_3 (0.1 ppm) and Na_2O (0.05 ppm) although it contains chlorine and hydroxyl group impurities at the level of 100 ppm and 1000 ppm, respectively. Type-IV silica is also synthesized from $SiCl_4$ by the oxidation with an oxygen plasma flame. It is similar to Type-III silica but the level of impurity content is much lower than the other (Al_2O_3 < 0.01 ppm, Na_2 < 0.02 ppm). Its hydroxyl content and Cl content are 0.4 ppm and 200 ppm, respectively. Type-V silica is synthesized from $SiCl_4$ or SiH_4 by reacting the raw material with H_2O using the chemical vapor deposition process under a flow of Ar and O_2 gas. The impurity content is the lowest among the silica glass produced by the existing techniques [24].

In addition to the above five types, a sol-gel-derived silica is receiving attention, although there are various issues to overcome before it can be brought into commercial application. A sol-gel-derived silica glass contains neither oxide nor chlorine impurity but does contain water. The glass shows a good transparency in the UV region [50].

1.3.1.2 Silicate glasses

The reduction of melting temperature from that of silica (about 2000 °C) to a level of 1500 °C that is more suitable to mass production is made by adding various modifier oxides to the glass composition. The addition of modifier oxides to silica leads to the breaking of the strong Si–O–Si bonds to yield weak points in the glass network [3], and thus reduces the melt viscosity, as it is schematically expressed by Eq. (1.17) using Na_2O as the modifier.

$$\equiv\!Si–O–Si\!\equiv\ +\ Na_2O\ =\ \equiv\!Si–O–Na\ Na–O–Si\!\equiv \qquad (1.17)$$

The oxygen ion bridging two silicon atoms in the left-hand side of the equation is called 'bridging oxygen', whereas the oxygen between Na and Si is called 'nonbridging oxygen'.

The representative of multicomponent silicate glass firstly developed by this concept is soda-lime silicate which is the most popular and makes up more than 90% of the commercial glasses. The typical composition of the glass is (70 $\sim$ 74)SiO_2–(12 $\sim$ 16)Na_2O–(8 $\sim$ 13)CaO–(1 $\sim$ 4)MgO–(0.5 $\sim$ 2.5)Al_2O_3 (wt%). The role of Na_2O is obviously to yield nonbridging oxygen and hence

Table 1.8. *Some properties of silica glass and representative commercial silicate glasses*

Glass	T_s (°C)	T_A (°C)	T_f (°C)	α (10^{-7}/°C)	d (g/cm^3)	n_d	log (Ω cm)
Silica glass	1070	1140	1667	5.5	2.20	1.4585	12.5
Sheet glass by float	510	553	735	87	2.50		
Container glass	505	548	730	85	2.49	1.520	7.0
Crystal glass	417	454	642	99	2.98	1.565	–
Optical glass SF6	–	423	538	90	5.18	1.80518	–
Optical glass BK7	–	559	719	83	2.51	1.51680	–
Laboratory ware	520	565	820	32	2.23	1.474	8.1
Alkali free glass	593	639	844	46	2.76	–	–
Electric glass (stem)	–	440	620	92	3.03	–	9.7
TV panel glass	473	514	704	101	2.77	–	–
TV funnel	450	490	672	100	3.01	–	–
TV neck	427	466	645	96.5	3.27	–	–

T_S (°C); Strain point, temperature at which the glass viscosity becomes $10^{14.5}$ dPa s
T_A (°C); Annealing point, temperature at which the glass viscosity becomes 10^{13} dPa s
T_f (°C); Softening point, temperature at which the glass viscosity becomes $10^{7.5}$–10^8 dPa s

to reduce the melt viscosity [51]. The addition of CaO (and MgO) is intended to improve the chemical durability of the glass which is degraded by the introduction of mobile sodium ions. The role of Al_2O_3, which effectively lowers the liquidus temperature of the system, is to improve the thermal stability and to hinder devitrification, as well as to improve the chemical durability. The biggest advantage of soda-lime silica glass over others is the low cost of raw materials and good workability which is essential for the fabrication of the products with a variety of shapes.

Replacement of Na_2O and/or CaO by other oxides such as K_2O, PbO, SrO, B_2O_3, etc. leads to the modification of the properties of glass. Some properties of the commercial silicate glasses detailed in Table 1.2 are shown in Table 1.8 together with the properties of silica glasses [52].

1.3.1.3 Nonsilicate oxide glasses

The special glasses used as key components of various devices in the fields of optics, electronics, and opto-electronics are not always silicates but are often nonsilicate glasses of the phosphate, borate, germanate, vanadate, or telluride systems. Although nonsilicate glasses are not generally applied to mass production due to the high cost of raw materials and their rather inferior chemical durability, they do show unique properties that cannot be obtained for

Table 1.9. *Solubility of silver in various oxide glasses*

Glass (mol%)	Solubility (Ag$_2$O mol%)	Ref.
SiO$_2$	−0	[54]
GeO$_2$	−25	[55]
B$_2$O$_3$	−33	[56, 57]
P$_2$O$_5$	−66	[58]
20Na$_2$O-80SiO$_2$	2–3	
20Na$_2$O-80GeO$_2$	2–3	
20Na$_2$O-80B$_2$O$_3$	11–12	
20Na$_2$O-80P$_2$O$_5$	54–57	

silicate glasses. The formation of these nonsilicate oxide glasses is also made by melt-quenching technique but replacing a tank furnace by a proper crucible made of materials such as platinum, platinum/rhodium, or gold which do not react with melts and therefore do not contaminate the products.

Multi-component phosphate glasses primarily consist of P-O-P chains of four-fold coordinated phosphorus [51]. The features of phosphate glasses as materials for optical application include a lower nonlinear refractive index and a high absorption cross-section [53], which is more suitable than silicate as the host glass for giant pulse lasers. The chemical composition of represent-ative glasses used in this application is $(61 \sim 66)$P$_2$O$_5$–$(4 \sim 8)$Al$_2$O$_3$–$(11 \sim 14)$K$_2$O–$(5 \sim 16)$BaO–$(0 \sim 1)$Nb$_2$O$_5$–$(1 \sim 3)$Nd$_2$O$_3$ (wt%). Their relevant properties are reviewed in detail in Chapter 3.

Table 1.9 shows the solubility of silver in various oxide glasses as an example [54–58], and also demonstrates that phosphate glasses show the high solubility of novel metals and nonoxide semiconductor compounds such as CdS, CdSe, etc., that is another of its advantages over silicate in the preparation of glass with special properties. The gradient-refractive-index materials, and the noble metal or semiconductor-doped nonlinear optical materials of high susceptibility based on these features are reviewed in Chapters 2 and 4 respectively.

Phosphate glasses have a lower softening point than silicate glasses and have long been used as the enamel for aluminum metal and as solder glasses for glass-to-glass bonding [59]. Some examples of phosphate glasses used in this application are shown in Table 1.10 along with their properties.

Multi-component borate glasses which consist primarily of BO$_3$ triangles and BO$_4$ tetrahedra are formed by combination with alkali, alkaline earth oxides and alumina, etc. [51]. The most important feature of borate glasses for

Table 1.10. *Example of phosphate solder glasses and*
their properties [59b]

ZnO	PbO (mol%)	P_2O_5	Expansion coefficient α ($\times 10^{-7}$ °C)	Softening point T_f (°C)
70	30	–	131	300
50	50	–	171	310
50	–	50	84.5	380
50	30	20	121	325
50	20	30	116	330

Table 1.11. *Example of borate solder glasses and their properties*

B_2O_3	PbO	PbF_2	ZnO (wt%)	SiO_2	CuO	Al_2O_3	α ($\times 10^{-7}$ °C)	T_f (°C)	Ref.
9.0	84.0	–	7.0	–	–	–	117	333.8	[59c]
9.1	80.4	3.6	7.0	–	–	–	119	317.3	[59c]
9.2	73.2	10.8	7.0	–	–	–	130	300.3	[59c]
20	17	–	40	13	10	–	50.2	571	[59d]
29.1	–	–	–	–	51.9	19.0	35.2	570	[59e]
38.3	–	–	–	–	43.8	17.9	36.0	665	[59e]
47.2	–	–	–	–	43.8	9.0	37.5	650	[59e]

application in the optical and opto-electronic fields is their high compatibility with rare-earth elements. The representative example of material based on this feature is the La_2O_3-containing high index and low dispersion optical glasses [21] (e.g.; LaK10: $41.3B_2O_3$–$12.1CaO$–$6.1PbO$–$32.4La_2O_3$–$8.1ZrO_2$–$0.2As_2O_3$ (wt%), $n_d = 1.7200$, $v_d = 50.03$). The very specific application of this feature is a glass for the Faraday rotator [60] which is reviewed in detail in Chapter 5.

The glasses of the borate system have also been used widely as solder glasses for glass-to-glass and glass-to-ceramic bonding because of their low softening point [59]. Examples of borate glasses used in this application are shown in Table 1.11 along with their properties.

GeO_2 single-component glass consists of a random three-dimensional network of Ge–O–Ge bonds similarly to silica glass [51]. GeO_2 glass and multi-component germanate glasses show fair chemical durability but are seldom used practically as, although they have similar properties to those of silicate glasses the raw material is far more expensive. The only very special application developed for glass of this system is infrared transmitting glass [61]. The higher infrared transmission of germanate glass than silica is attracting people to use it as the potential material for ultra-low-loss optical fiber [62].

In addition to these representative glass-forming oxides, TeO_2 and V_2O_5 are also used for the preparation of glasses with special properties. A telluride glass is receiving attention as it is a nonlinear optical glass with ultra-fast response owing to its high refractive index [63]. V_2O_5 combined with P_2O_5 (and BaO) is well known as forming an electronic conducting semiconductor glass having resistivity of the order of $10^5 \sim 10^6 \, \Omega \, cm$ at room temperature [64]. The typical composition of these glasses are $(60 \sim 90)V_2O_5-(10 \sim 40)P_2O_5$, $60V_2O_5-xBaO-(40-x)P_2O_5$ (mol%), etc.

1.3.2 Halide and oxy-halide glasses

1.3.2.1 Fluoride glasses

Before the discovery of glasses of the $ZrF_4-BaF_2-NaF-NdF_3$ system in 1974 by Poulain and Lucas [65], BeF_2-based glasses were the only halide glass that could be applied practically. A BeF_2-based fluoride glass has the lowest refractive index and highest Abbe number of any inorganic glass-forming system [66], (e.g. $n_d = 1.3741$, $\nu_d = 105$ for $26.0BeF_2-18.0AlF_3-27.2CaF_2-29.7NaF$ (wt%) glass). It gives the least distortion for a high intensity optical beam due to its very low nonlinear refractive index. However, the glass is highly toxic and extremely hygroscopic and therefore its practical application is limited and little research into it has been done since the 1960s.

In contrast to fluoroberyllate glass, the heavy metal halide glasses represented by ZrF_4-based multicomponent fluoride glass are neither toxic nor hygroscopic. The chemical bond in these glasses is ionic in nature, and therefore, the glass shows a good transparency in the infrared region, as seen in Fig. 1.9, and this is advantageous in various optical fields [13].

The fluorozirconate glass represented by so called ZBLAN glass, which has the composition $56ZrF_4-19BaF_2-6LaF_3-4AlF_3-15NaF$ (atomic%), is transparent up to about 8 μm and has chemical durability which is comparable with a sodium silicate glass. Its theoretical optical loss was reported to be about $10^{-3} \, dB \, km^{-1}$ in the vicinity of $3 \sim 4$ μm by the NNT Japan group [67]. This value is about two orders of magnitude lower than the theoretical loss of silica glass. Thus glasses from this heavy metal fluoride system attracted people as the materials for the long distance fiber optical waveguides for transoceanic or transcontinental communication systems, which led to the development of a wide variety of multicomponent halide glasses by researchers all over the world

Table 1.12. *Chemical composition of representative heavy metal fluoride glasses and their properties*

Glass	Composition (mol%)	T_g (°C)	d (g/cm^3)	n_d	R_c (K/min)	Ref.
ZBLAN	$56ZrF_4-19BaF_2-6LaF_3-4AlF_3-15NaF$	269			3	[69]
ZBLAN	$53ZrF_4-20BaF_2-4LaF_3.3AlF_3-15NaF$	256	4.319	1.4985	1.7	[69]
ZBGA	$57ZrF_4-35BaF_2-5GdF_3-3AlF_3$				70	[70]
ZBLA	$57ZrF_4-36BaF_2-3LaF_3-4AlF_3$	310	4.61	1.516	55	[71]
HBLA	$61HfF_4-29BaF_2-7LaF_3-3AlF_3$				170	[71]
YABC	$20YF_3-40AlF_3-20BaF_2-20CaF_2$	430		1.43	60	[72]
CLAP	$26.1CdF_2-10LiF-30.6AlF_3-33.3PbF_2$	245	5.886	1.606	300	[73]

[68]. Examples of the composition of glasses based on heavy metal fluoride such as ZrF_4, HfF_4, YF_3, PbF_2, ZnF_2 are shown in Table 1.12 along with some of their properties.

The high transparency of these heavy metal fluoride glasses up to the mid-infrared region is a major advantage over the oxide glasses as host materials for various rare-earth doped lasers [68]. A wide variety of lasers to be used in the visible and near infrared regions, as well as up-conversion lasers, have been developed and practically applied in various fields, as reviewed in Chapter 3.

One important fact to be noted concerning halide glasses is that their preparation must be carried out in an oxygen-free atmosphere so that the reaction of the halide raw materials with oxygen which leads to the formation of oxide impurities is prevented [13]. The formation of the heavy metal fluoride glass is, therefore, usually made within a glove-box in a nitrogen atmosphere using a gold, platinum, or graphite crucible.

The drawback of glasses of heavy metal fluoride systems is the lower viscosity of melt compared with those of oxide systems [13]. It often precipitates crystals during the cooling of a melt or during re-heating of the glass for secondary processing such as fiber-drawing. Thus, the loss of fiber drawn from heavy metal fluoride glasses still remains around 1 dB km^{-1}, in spite of tremendous efforts by many researchers seeking to improve the thermal stability [68]. The modification of glass composition to improve the thermal stability has been carried out using 'critical cooling rate' as a parameter representing the thermal stability of a melt during cooling [69].

The 'critical cooling rate' is defined as the slowest cooling rate at which a glass is obtained without crystallization from a melt of given composition. It is empirically correlated with the onset temperatures of the exothermic peak, T_{xc}, in the DTA curves obtained for a melt at various cooling rates by Eq. (1.18).

$$\ln(R) = \ln(R_{\mathrm{c}}) - 1/(T_1 - T_{\mathrm{xc}})^2 \qquad (1.18)$$

where R_{c} is critical cooling rate, T_1 is the liquidus temperature of the melt, and T_{xc} is the onset temperature of the exothermic peak for the cooling rate, R. In practice, R_{c} was determined by extrapolating the logarithm of cooling rate R plotted against the inverse of the square of $(T_1 - T_{\mathrm{xc}})$ for the range $R = 1 \sim 100 \, \mathrm{K\,min^{-1}}$ as shown in Fig. 1.17. The critical cooling rate, R_{c}, of ZBLAN and YABC glasses which show high thermal stability among the heavy metal fluoride glasses are about 0.7 and 20 $\mathrm{K\,min^{-1}}$ [74], respectively, whereas that of silica glass is in the order of $10^{-6} \, \mathrm{K\,min^{-1}}$ [9].

Another important parameter obtained from the DTA curve and used for the modification of glass composition is the difference between the onset temperature of the exothermic peak during heating, T_{xh}, and the glass transition point T_{g}, i.e. $\Delta T = (T_{\mathrm{xh}} - T_{\mathrm{g}})$ or normalized value of ΔT by T_{g}, i.e., $(T_{\mathrm{xh}} - T_{\mathrm{g}})/T_{\mathrm{g}}$ [13]. The value of ΔT or $\Delta T/T_{\mathrm{g}}$ represents the stability of the glass against reheating for the secondary treatment. The larger the ΔT, the higher the stability.

The modification of glass composition is made by trial and error so that the lower value of R_{c} and the larger value of $\Delta T/T_{\mathrm{g}}$ are obtained for the system of interest. Figure 1.18 shows an example of such experiments to modify the

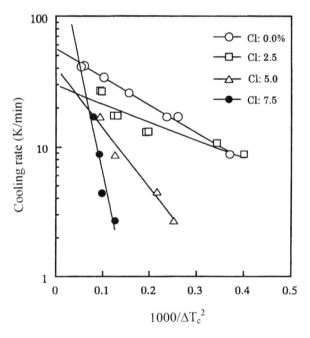

Fig. 1.17. Plot of cooling rate against $1/(T_1 - T_{\mathrm{xc}})^2$ on the melt of composition $20YF_3-40AlF_3-20BaF_2-20CaF_2$. [Reprinted from T. Yano, J. Mizuno, S. Shibata, M. Yamane, S. Inoue and Y. Onoda, *J. Non-Cryst. Solids*, **213 & 214** (1997) 345; copyright (1997) with permission from Elsevier Science.]

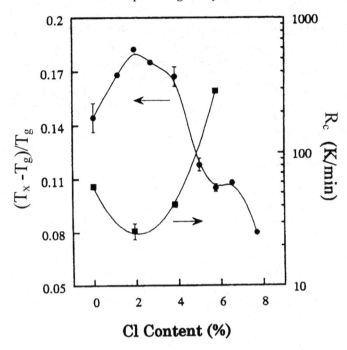

Fig. 1.18.   Improvement of thermal stability of $20YF_3-40AlF_3-20BaF_2-20CaF_2$ glass with the addition of $CaCl_2$. [Reprinted from T. Yano, J. Mizuno, S. Shibata, M. Yamane, S. Inoue, and Y. Onoda, *J. Non-Cryst. Solids*, **213 & 214** (1997) 345, copyright (1997) with permission from Elsevier Science.]

thermal stability of a glass of $YF_3-AlF_3-BaF_2-CaF_2$ system with the addition of $CaCl_2$ [74]. It can be seen from the figure that the thermal stability of the glass is improved by the addition of $CaCl_2$ up to about 2 mol%.

An excellent review on the preparation, properties and application of heavy metal fluoride glass is found in the book by P. W. France, M. G. Drexhage, J. M. Parker, M. W. Moore, S. F. Carter, and J. V. Wright [68a].

1.3.2.2 Oxy-halide glasses

There are several important glasses prepared from the batch mixture of oxides and halides. One of these is fluorophosphate glass of the system $P_2O_5-AlF_3-MgF_2-CaF_2-SrF_2-BaF_2$. Glasses of this system have low dispersion and a low nonlinear refractive index, and are used as host materials for high-power laser systems [75] as reviewed in Chapter 3.

The other example of oxy-halide glass is in the system $AgI-Ag_2O-MoO_3$ (or P_2O_5, etc.) [76]. The glasses of this system consist of discrete ions of I^-, Ag^+, and MoO_4^{2-} (or PO_4^{3-}), and display the characteristics of a solid-state electrolyte with a very high ionic conductivity of the order of 10^{-2} S cm^{-1} at room

Table 1.13. *Example of glasses of AgI–Ag$_2$O–MoO$_3$ (or P$_2$O$_5$, B$_2$O$_3$) systems and their ionic conductivity [76]*

Glass composition (mol%)	Conductivity (S cm^{-1})	Transport number of Ag$^+$
75AgI–25Ag$_2$MoO$_4$	1.1×10^{-2}	0.999_6
60AgI–40Ag$_2$MoO$_4$	6.0×10^{-4}	0.999_6
80AgI–20Ag$_3$PO$_4$	1.2×10^{-2}	0.997_2
60AgI–30Ag$_2$O–10B$_2$O$_3$	8.5×10^{-3}	0.992_7
60AgI–40Ag$_2$SeO$_4$	3.1×10^{-3}	0.998_8
50AgI–30Ag$_2$O–20V$_2$O$_5$	1.3×10^{-3}	0.995_9
33AgI–33Ag$_2$O–33GeO$_2$	1.3×10^{-4}	0.985_3

Table 1.14. *Example of low melting glasses of PbF$_2$–SnF$_2$–P$_2$O$_5$ system and their properties [77]*

Sn	Pb	P	F	O	T_g (°C)
49.5	3.0	47.5	45.7	146.4	124
53.4	3.8	42.8	53.5	137.5	120
56.6	4.4	39.0	62.1	127.5	87
61.0	5.2	33.8	73.8	113.8	85
50.5	5.7	38.2	38.2	146.7	135
50.0	5.5	44.4	35.3	148.9	145

Glass composition is expressed so that the sum of cation becomes 100, i.e. Sn + Pb + P = 100 in atomic ratio.

temperature as shown in Table 1.13. Although they are not thermally stable and must be prepared by rapid quenching using a twin-roller device, the high ionic conductivity is attractive as an electrolyte for an ionic battery.

Other examples of oxy-halide glasses with special properties are those in the system PbF$_2$–SnF$_2$–P$_2$O$_5$ shown in Table 1.14 [77]. The glasses of this system are melted at a temperature around $400 \sim 500$ °C. In spite of its low glass transition temperature, in the vicinity of $90 \sim 120$ °C, the glass shows fair chemical durability that is acceptable, depending on the application. The glass is considered to contain a fairly large number of less polar P–F bonds and has the structure of a molecule in nature [78]. Because of this unique structure and a very low melting temperature, the glass can incorporate active organic dye molecules such as Rhodamine 6G, Phtalocyanine, Fluorescein, etc. without deteriorating the activity of the molecules. Thus, a potential use of the glass has been suggested for solid-state dye lasers.

Table 1.15. *Composition and properties of representative chalcogenide glasses*
[80, 81]

Glass	n	λ (µm)	T_f (°C)	α ($\times 10^7$/°C)	ρ (Ω cm, 130 °C)
As_2S_3	2.35	$1.5 \sim 12$	180	300	2.0×10^{12}
As_2Se_3		$1 \sim 15$	194		1.54×10^8
As_2Se_2Te					2.5×10^5
As_2SeTe_2					3.5×10^3
$As_{12}Ge_{33}Se_{55}$	2.49	$1 \sim 16$	474		
$As_{20}S_{20}Se_{60}$	2.53	$1 \sim 13$	218	200	
$As_{50}Te_{10}S_{20}Se_{20}$	2.51	$1 \sim 13$	195	270	
As_8Se_{92}	2.48	$1 \sim 19$	70	340	
$As_{20}Te_{70}Ge_{10}$	3.55	$2 \sim 20$	178	180	
$Ge_{30}P_{10}S_{60}$	2.15	$2 \sim 8$	520	150	
$Ge_{40}S_{60}$	2.30	$0.9 \sim 12$	420	140	
$Ge_{28}Sb_{12}Se_{60}$	2.62	$1 \sim 15$	326	150	
$Si_{25}As_{25}Te_{50}$	2.93	$2 \sim 9$	317	130	

1.3.3 Chalcogenide glasses

Chalcogenide glasses are based on sulfur, selenium, or tellurium and are obtained by melting these elements or their compounds with As, Sb, Ge, Si, etc. in an oxygen-free atmosphere, i.e. in vacuum or within a silica glass ampoule, so that reaction of the raw materials with oxygen is avoided [79].

These glasses differ from the glasses of oxide and halide systems mentioned above as the atomic ratio between the positive elements such as As, Ge, Si, etc. and the negative elements such as S, Se, Te, etc. is not always constant in chalcogenide glasses but can be varied within a certain range. In the case of the As–S system, for example, glass formation is possible if the atomic ratio As/S is within the range of $1.5 \sim 9$ [79]. Thus the composition of chalcogenide glasses is usually expressed using the atomic percentage or atomic ratio of the constituent elements in such a way as $Ge_{33}As_{12}Se_{55}$, $Si_{25}As_{25}Te_{50}$, $Ge_{30}P_{10}S_{60}$, As_2Se_3, $Ge_3P_3Te_{14}$, etc.

The chemical bonds in the chalcogenide glasses are covalent in nature [79], in contrast to the ionic nature of bonds in oxide and halide glasses. One of the most important features of chalcogenide glasses is good transparency in the infrared region up to about 20 µm as already shown in Fig. 1.9 using As_2Se_3 and $Ge_{10}As_{50}Te_{40}$ glass as examples. Thus, the glasses of this system have been used widely as optical elements for infrared rays such as prisms, windows, waveguides, etc., as well as the host materials of infrared lasers [80], which is reviewed in Chapter 3 in detail.

Another feature of chalcogenide glasses is that they are band-type semiconductors [81] and show various unique phenomena such as photoluminescence (As_2Se_3) [82], photoconduction (As_2Se_3, As_2Te_3, $Tl_2SeAs_2Te_3$, etc.) [83], photodarkening ($Te_{81}Ge_{15}Sb_2S_2$) [84], memory-switch ($Te_{81}Ge_{15}Sb_2S_2$) and threshold-switch ($As_{30}Te_{48}Ge_{10}Si_{12}$, $As_{43}Te_{21}Tl_{14}Sb_6Se_{16}$, etc.) [85] in voltage-current relationships, etc. Applications of these features in the field of optics and opto-electronics include photoconductive film for xerography, resistance material for photolithography, etc. Composition and properties of representative chalcogenide glasses are shown in Table 1.15.

1.4 Secondary treatment of a glass for the development of a special function

1.4.1 Thermal treatment

1.4.1.1 Crystallization

It was mentioned in Section 1.1.2 that a glass is a thermodynamically metastable material and remains untransformed to the most stable state, i.e. crystalline state, due to a hindrance of atomic rearrangement. Therefore, the crystallization of a glass can take place if the atomic rearrangement is enhanced by reheating to a temperature above T_g. If the thermal treatment for the crystallization of a glass is carried out so that the precipitation of crystallites of unique properties occurs throughout the material, the development of new functionality is possible.

Materials prepared by the crystallization of a glass with the purpose of developing this new functionality are called glass ceramics [8]. The properties and characteristics of crystallites may be a low (or zero) thermal expansion coefficient [86], high mechanical strength, low dielectric constants [87], high electric resistivity, good machinability [88], biological-activity [89], etc. Table 1.16 shows some examples of glass ceramics.

The advantage of glass ceramics over ordinary ceramics obtained by the sintering of powdery raw materials is that the thermal treatment for crystallization is made, in general, on a glass that has been formed in advance into the desired shape, and therefore, a pore-free material of complicated shape can be obtained.

As an industrial material is always required to have high mechanical strength, and the mechanical strength of glass ceramics is dependent on the size of the precipitating crystal, the heat treatment for nucleation which determines the micro-structure of the eventual glass ceramics is one of the most important steps in the preparation process. As discussed in Section 1.1.2,

Table 1.16. *Examples of important glass ceramics*

System	Main crystalline	Feature	Application	Ref.
A) $Li_2O-Al_2O_3-SiO_2$ $(TiO_2, ZrO_2, P_2O_5)^*$	β-spodumene ss	Low expansion, (translucent)	Table ware, Oven top, Work table top	[86a]
B) $Li_2O-Al_2O_3-SiO_2$ $(TiO_2, ZrO_2, P_2O_5)^*$	β-quartz ss, β-spodumene ss	Low expansion, (transparent)	Astronomical mirror Laser device	[86b, 86c]
C) $Li_2O-Al_2O_3-SiO_2$ $(Au-Ce, Ag-Ce)$	Li_2O-SiO_2, $Li_2O-2SiO_2$	Photosensitive, chemical machining	Display cell sheet, Fluidic device	[86d]
D) $MgO-Al_2O_3-SiO_2$ (TiO_2)	cordierite	High strength, Low dielectric const.	Radar antennae Large electric insulator	[87]
E) $K_2O-MgO-Al_2O_3-SiO_2-F_2$	mica (fluorophlogopite)	Machinable	Hermetic joint Vacuum equipment	[88]
F) $Na_2O-K_2O-BaO-ZnO-CaO-B_2O_3-Al_2O_3-SiO_2$	wollastonite	Appearance of marble	Interior application Exterior application	[90]
G) $CaO-MgO-SiO_2-P_2O_5-CaF_2$	apatite, wollastonite	Bioactive	Replace bone tissue	[89]

* nucleating agent

both nucleation rate I_0 and crystal growth rate U are dependent on the temperature of the heat treatment and expressed by Eqs (1.2) and (1.3),

$$I_0 = k_1/\eta \exp(-g\alpha^3\beta/T_r(\Delta T_r)^2) \tag{1.2}$$

$$U = f(RT/3N_A\pi\lambda^2\eta)(1 - \exp[\Delta H_{fM}\Delta T/RTTm]) \tag{1.3}$$

The maximum I_0 appears at a lower temperature than the maximum U as shown in Fig. 1.5. The most effective heat treatment for nucleation is obviously attained at the temperature that gives the maximum nucleation rate. Therefore, a glass that has been cooled down to room temperature after forming is generally re-heated to this temperature, followed by treatment at a high temperature for crystal growth [8], as is schematically shown in Fig. 1.19.

The optimum temperatures for these treatments are obviously dependent on glass composition and must be experimentally determined. The optimum temperature for crystal growth is considered to be in a range slightly above the temperature of the exothermic peak T_{p0} of untreated glass. To find the optimum temperature for nucleation, a sample glass is first subjected to heat treatment for a predetermined time, e.g. 2 h, at a temperature, T_1, between the glass transition temperature T_g and the onset temperature T_{xh} of the exothermic peak in DTA [91]. The treated glass is then pulverized and subjected to DTA analysis. The temperature of the exothermic peak T_{p1} will be observed at a temperature that is lower than T_{p0}, if the heat-treatment is sufficient, as shown in Fig. 1.20.

By repeating this procedure at various temperatures, T_1, T_2, T_3, $\ldots$, and plotting the reciprocal of the difference between T_{p0} and T_{pi} against respective

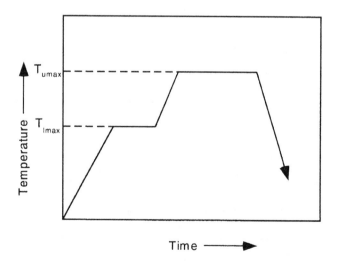

Fig. 1.19.   Schematic illustration of heat cycle for the preparation of glass ceramics

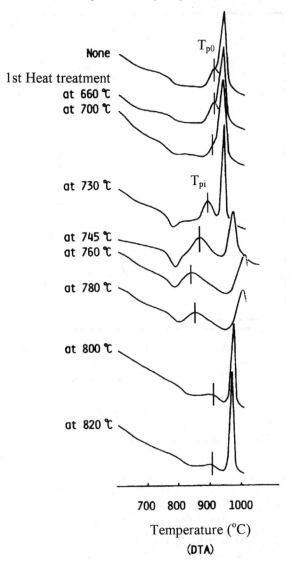

None

1st Heat treatment
at 660 ℃
at 700 ℃

at 730 ℃

at 745 ℃
at 760 ℃

at 780 ℃

at 800 ℃

at 820 ℃

T_{p0}

T_{pi}

700   800   900   1000

Temperature (°C)

(DTA)

Fig. 1.20.   Shift of the temperature of exothermic peak toward lower temperature by the treatment for nucleation. [Reprinted from X. Zhou and M. Yamane, *J. Ceram. Soc. Jpn.* **96** (1988) 152, copyright (1988) with permission from the Ceramic Society of Japan.]

temperatures, we obtain a curve similar to the one showing the temperature dependence of the nucleation rate [91] as shown in Fig. 1.21.

The pictures in Fig. 1.22 are electron micrographs of the fractured surface of samples obtained by the heat treatment for nucleation at various temperatures of T_1, T_2, T_3, ..., followed by the treatment for crystal growth at 850 °C [91]. It is known that the crystal size is the smallest in the sample obtained by heat

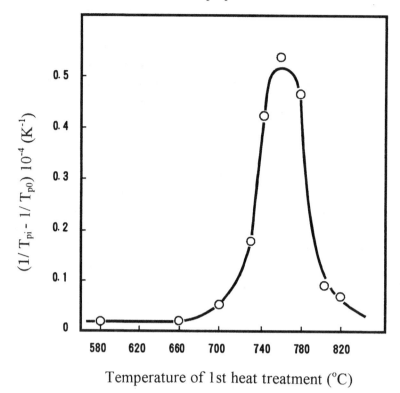

Fig. 1.21. Plot of the reciprocal of the difference between T_{p0} and T_{pi} against temperature of heat treatment for nucleation. [Reprinted from X. Zhou and M. Yamane, *J. Ceram. Soc. Jpn.* **96** (1988) 152, copyright (1988) with permission from the Ceramic Society of Japan.]

treatment for nucleation at 760 °C at which the value of $\Delta T_p (= T_{p0} - T_{pi})$ becomes the maximum, showing the validity of the method.

Another important parameter that must be taken into consideration is the kind and the amount of nucleating agent [8]. The role of the nucleating agent is to provide foreign surfaces within a glass by precipitating metal colloids or by inducing micro-phase separation so that the heterogeneous nucleation throughout the bulk of the glass is enhanced and surface crystallization which often degrades the quality is hindered. The familiar nucleating agents used for the preparation of glass ceramics shown in Table 1.16 include novel metals such as Au, Ag, Pt, etc. for colloid formation and TiO_2, P_2O_5, ZrO_2, etc. for inducing the phase separation of mother glasses.

The convenient way to determine the optimum amount of these nucleating agents is to carry out a DTA analysis on the pulverized sample glass of two different sizes, i.e. about 100 μm and 20 μm on average [8]. If the heterogeneous nucleation occurs ideally throughout the bulk glass without surface

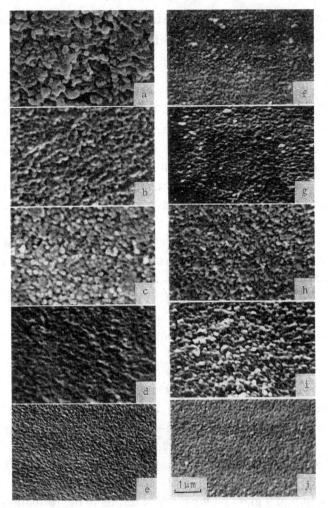

Fig. 1.22. Electron micrographs of the fractured surface of samples obtained by the heat treatment for 2 h for nucleation at various temperatures (a: none; b: 580 °C; c: 700 °C; d: 730 °C; e: 745 °C; f: 760 °C; g: 780 °C; h: 800 °C; i: 820 °C; j: 760 °C (16 h)) and subsequent heating at 850 °C for crystal growth. [Reprinted from X. Zhou and M. Yamane, *J. Ceram. Soc. Jpn.* **96** (1988) 152, copyright (1988) with permission from The Ceramic Society of Japan.]

nucleation, the exothermic peak is observed at the same temperature independently of the size of the glass sample. On the other hand, if the nucleating agent is not effective, the influence of surface nucleation will be reflected to the exothermic peak, and the peak for the sample of larger size will appear at a higher temperature than that for the sample of smaller size due to the smaller specific surface area than the latter.

Excellent reviews of the theory of crystallization and practical glass ceramics are found in the book edited by J. H. Simmons, D. R. Uhlmann and G. H. Beall [92], and the book by Z. Strunad [8].

1.4.1.2 Precipitation of nano-particles

As mentioned in the previous section, one of the advantages of a glass over ceramic materials is the flexibility of chemical composition which enables the doping of particular ions or active elements for the development of special color, emission of fluorescent light, high nonlinear susceptibility, photochromism, etc. Glasses of various colors are obtained by doping transition metal ions, rare-earth elements, noble metals and semiconductor compounds [93], as shown in Table 1.17.

Among those in the table, the colors by Au, Ag, Se, CdS, CdSe, etc. are obtained from their colloids which are up to 10 nm in diameter. Glasses containing these colloid particles also display high nonlinear susceptibility (Au, Ag, CdS, CdSe, CuCl colloid) [94], or photochromism (AgCl, TlCl) [95]. The performance of a glass in terms of these properties is dependent on number density and on the size and size distribution of the colloid particles as reviewed in Chapter 4.

Since the size and size distribution of colloids are dependent on the thermal history of a glass [93], the precisely controlled heat treatment is as important as the composition in order to develop a glass with special function. The precipitation of colloids of metals, non-oxide semiconductor compounds and silver halides, in general, proceeds according to the nucleation and growth mechanism [93]. The amount of precipitating colloids depends on the solubility of constituents in a glass melt at high temperature and its temperature coefficient. Slow cooling of a glass melt generally results in the formation of colloids with a wide size distribution, particularly when the amount of precursor ions is large. Therefore, a glass melt is usually quenched rapidly and re-heated to a temperature above T_g to enhance the nucleation and growth of particles so that a glass that contains nano-particles of uniform size distribution is obtained.

In the case of colloid formation of gold [93], for example, the atomic gold forms nuclei during the cooling period. The nuclei develop into the color centers when heated later. Rapid cooling produces a high degree of super-saturation which leads to the desirable large number of invisible nuclei $2 \sim 4$ nm in size and consisting of several hundred atoms. On reheating the glass, gold atoms which had been 'frozen in' can migrate within the glass and collide with the nuclei, resulting in the growth of the nuclei into larger crystals. By raising the temperature above the softening point, small crystals can move and collide, leading to the formation of secondary particles (clusters). Further

Table 1.17. *Color of glass that has been doped with various ions and elements [93]*

Element	Chemical state	Color
V	V^{5+}	Faint yellow
	V^{4+}	Blue
	V^{3+}	Green
Cr	Cr^{6+}	Yellow
	Cr^{3+}	Green
Ti	Ti^{3+}	Violet-purple
Mn	Mn^{3+}	Purple
	Mn^{2+}	Light yellow
Fe	Fe^{3+}	Faint yellow
	Fe^{2+}	Blue-green
Co	Co^{3+}	Faint pink
	Co^{2+}	Deep blue
Ni	Ni^{2+}	Brown-purple
Cu	Cu^{2+}	Blue-green
	Cu colloid	Red
Ce	Ce^{4+}	Yellow
Pr	Pr^{3+}	Green
Nd	Nd^{3+}	Purple
Sm	Sm^{3+}	Yellow
Eu	Eu^{2+}	Yellow
Ho	Ho^{3+}	Yellow
Er	Er^{3+}	Pink
Ag	Ag colloid	Yellow
Au	Au colloid	Red
Se	Se colloid	Pink
P	P colloid	Red
Te	Te colloid	Red
Cd, S	CdS colloid	Yellow
Cd, Se	CdSe colloid	Pink
Cd, S, Se	CdS_xSe_{1-x} colloid	Orange
Fe, S	FeS colloid	Amber
Sb, Se	Sb_2Se_3 colloid	Red

heating causes the recrystallization leading to a reduced number of larger crystals at the expense of smaller crystals.

Table 1.18 shows the color change of gold ruby plate glass with temperature and time of heat treatment [93]. At 575 °C, the glass develops a faint purple (amethyst color) after 2 h, prolonged heating up to 9 h makes the color more intense and a bluish purple color giving an absorption maximum at 560 nm becomes evident. At 600 °C, the purple color develops much faster than at 575 °C in the first few hours. The prolonged heating up to 9 h results in a shift

Table 1.18. *Color change of gold ruby plate glass with*
temperature and time of heat treatment [93a]

Temperature (°C)	2 h	Time 4.5 h	9 h
575	colorless	bluish-purple	bluish-purple
600	purple	purple	ruby
675	intense ruby	intense ruby	intense ruby
700	blue*, brown**	blue*, brown**	blue*, brown**

*: transmitting light; **: reflected light.

of maximum wavelength from 560 nm to 532 nm and gives the real ruby color due to recrystallization. Above 600 °C, the effect of recrystallization becomes evident, and at 675 °C the ruby color is fully developed in less than 2 h. The prolonged heating at this temperature does not have a great effect on color and intensity, despite the fact that the recrystallization proceeds (maximum wavelength remains at 550 nm). At 700 °C, some of the gold crystals grow at the expense of smaller ones. The larger crystals bring about reflection and scattering, and give blue color in transmitted and brown color in reflected light (maximum wavelength shifts to 560–570 nm), showing the excessive crystal growth. Thus, 675 °C is the most desirable temperature for the development of ruby color for soda-lime silicate glass. In the case of sodium borosilicate glass of $75.0SiO_2 - 12.0B_2O_3 - 3.0Al_2O_3 - 6.0Na_2O - 4.0Sb_2O_3$ (wt%), on the other hand, the heat treatment for the development of the gold ruby color is carried out at a temperature around 800 °C for 10 min or a little longer [93].

The development of color or colloid formation for other metals and semiconductor compounds occurs in a similar manner to this case, although the suitable temperatures and times of the treatment differ from system to system and must be experimentally determined case by case.

1.4.1.3 Phase separation

Another important thermal treatment is enhancement of glass-glass phase separation. In some systems such as Na_2O-SiO_2, Li_2O-SiO_2, $Na_2O-B_2O_3-SiO_2$, $Li_2O-BaO-SiO_2$, etc., there are composition regions in which a glass shows metastable immiscibility [96]. A glass of the Na_2O-SiO_2 system, for example, is single phase at a temperature above the solid curve shown in Fig. 1.23, but decomposes into two glassy phases when cooled to a temperature range below the curve.

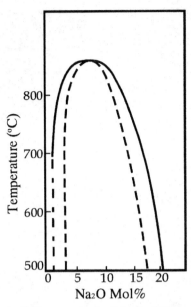

Fig. 1.23. Metastable immiscibility cupola for Na_2O-SiO_2 system. [Reprinted from V. N. Filipovich, *Phase Separation in Glass*, ed. O. V. Nazurin & E. A. Pori-Koshits (Elsevier Science, 1984) p. 15, with permission of editors.]

There are two types of phase separation, spinodal decomposition and binodal decomposition, depending on the composition and thermal history of the glass. The phase separation by binodal decomposition proceeds in the regions between the solid and broken lines in Fig. 1.23 via nucleation and growth mechanisms and gives a microstructure that consists of dispersed microspherical glass particles in the matrix [96]. Spinodal decomposition proceeds in the region inside the broken curves via a mechanism with no nucleation process and gives the interconnected structure of glasses of two different compositions [96]. Figure 1.24 shows examples of the morphology of the microstructure of the phase-separated glasses produced by the respective mechanisms [97].

Phase separation differs from the crystallization of a glass as in this case the volume ratio of the two glass phases remains unchanged once it reaches an equilibrium value determined by the composition of the original glass and the temperature of treatment. However, the size and number of dispersed spheres or the size of two interconnected phases changes with time during isothermal treatment in such a way as to reduce the interfacial area between two phases.

As the eventual two glass phases differ in chemical composition, their chemical durabilities are also different. Thus, we can obtain a porous glass by leaching out one of the phases with acid. The representative of such a glass is

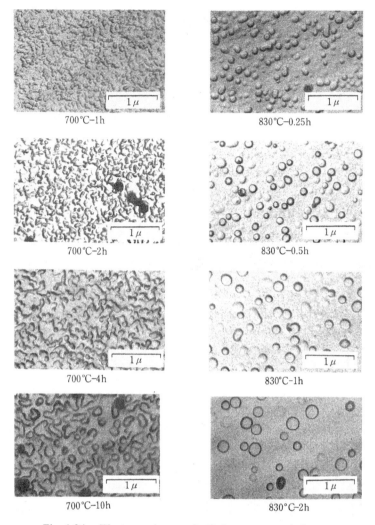

700°C–1h

830°C–0.25h

700°C–2h

830°C–0.5h

700°C–4h

830°C–1h

700°C–10h

830°C–2h

Fig. 1.24.   Electron micrograph of phase-separated glasses.

one which has the commercial name of Porous Vycor. It is obtained by inducing spinodal decomposition of a glass of composition about $70SiO_2-25B_2O_3-5Na_2O$ (mol%) and subsequently leaching the sodium–borate-rich phase to leave a skeleton of high silica glass of approximate composition $96SiO_2-4B_2O_3$ (wt%) [98]. The typical temperatures for glass melting and the heat treatment for phase separation are about 1400 °C and 500 ∼ 600 °C (4 ∼ 24 h), respectively. The typical conditions of acid treatment for leaching is for 24 ∼ 72 h at 95 °C with hydrochloric acid of 1 ∼ 3 N.

Some properties of Porous Vycor are shown in Table 1.19 along with those

Table 1.19. *Example of porous glasses and their properties*

Component	Vycor type [98]	PPG High silica [99]	Shirasu Glass [100]	TiO_2–SiO_2 system* [101]
SiO_2	55–80 (wt%) (96.3)	30–50 (wt%) (97–98)	49–50 (wt%) (69–70)	26–34 (mol%) (50–62)
B_2O_3	22–33 (2.9)	40–60 (1–2.5)	15–16 (6–7)	5–7 (<0.1)
Al_2O_3	0–2 (0.4)	0–4 (0–0.25)	9–10 (12–13)	10–16 (0.1–0.2)
Na_2O	7–10 (<0.2)	7–10 (<0.1)	4–5 (4–5)	–
K_2O	–	–	2–3 (3–4)	–
CaO	–	–	17–18 (2–3)	22–28 (0–0.3)
MgO	–	–	–0.2 (<0.1)	1.5–5 (0–0.2)
P_2O_5	–	–	–	5–7 (<0.1)
TiO_2	–	–	–	17–26 (37–50)
Pore volume	0.2–1.0 (cm^3/g)	0.4–0.6	0.4–0.6	0.3–0.5
Surface area	8–400 (m^2/g)	200–300	0.2–8	120–150
Pore diameter	0.5–300 (nm)	0.5–20	200–1000	4–20
Maximum Temp.	800 (°C)	850	600	950
Durability	stable to acid	stable to acid	stable to acid and base	stable to acid and base

* Glass ceramics.

of other porous glasses or porous glass ceramics based on phase separation. The Porous Vycor used to be produced as the precursor for a high silica glass which is obtained by sintering a porous glass into a pore-free one at a temperature around 1000 °C. The eventual glass had thermal properties that were comparable with silica glass and was used as an alternative to expensive silica glass [102]. After the development of the CVD process, however, Porous Vycor is seldom used as the precursor of silica glass, and is now used without sintering to make use of advantages inherent in the porous structure. Examples of applications of Porous Vycor in the preparation of glass with special properties are for the fabrication of GRIN rods and Slab lenses [103]. An excellent review on the theory and application of phase separation in glass is found in the book edited by O. V. Mazurin and E. A. Porai-Koshits [96].

1.4.2 Ion exchange

Mono-valent ions such as alkali, silver, copper, thallium, etc. in a glass are easily ionized by thermal energy and migrate within a glass from one stable position to another. Therefore, if a glass containing these cations is immersed in a medium containing other mono-valent ions such as a fused salt bath of alkali nitrate, chloride, etc., the exchange of ions in the glass with those in the bath takes place [104]. The ion exchange thus leads to the modification of the properties as the result of the change in chemical composition near the surface of the glass.

If alkali ions of small size are partially replaced by ions of a larger size, e.g. Na^+ by K^+, the mechanical strength of the eventual glass increases [105] from a few $kg\,mm^{-2}$ before the treatment to about $40\,kg\,mm^{-2}$ due to the development of compressive stress at the glass surface. If ions of high index-modifying ability are partially replaced by those of low index-modifying ability, e.g. Tl^+ by K^+, Ag^+ by Na^+, Li^+ by Na^+, etc., the eventual glass has a gradient in the refractive index which decreases from the center towards the periphery [106]. If ions of low mobility are totally replaced by those of high mobility, e.g., Na^+ by Cu^+ or Ag^+, the electrical conductivity of the eventual glass increases [107]. Thus the technique of ion exchange has been used in the development of gradient-index glass rods reviewed in Chapter 2, as well as the development of chemical-tempered glasses, optical waveguides, two-dimensional arrays of micro-lens [108], etc.

The rate of ion exchange depends on the mutual diffusion constants of ions which is given by Eq. (1.19) using self-diffusion constants of two ions A and B as [109].

$$D_M = (d \ln a_A / d \ln N_A) D_A D_B / (N_A D_A + N_B D_B) \qquad (1.19)$$

where, D_M is the mutual diffusion constant, D_A and D_B are the self-diffusion constants of ion A and B, N_A and N_B are the mole fraction of ion A and B and a_A is the activity of ion A, respectively. The self-diffusion constants of the alkali ions, D_A and D_B, are dependent on both glass composition and the concentrations of the alkali itself, and vary with temperature according to the Eq. (1.20),

$$D = D_0 \exp(-E_D / RT) \qquad (1.20)$$

where D_0 is a constant, E_D is the activation energy of diffusion and R is gas constant, respectively.

The values of D_0 and E_D for various ions in some silicate glasses are summarized in the handbook [110]. Some of the values related to the ion exchange for strengthing or GRIN perparation are shown in Table 1.20. By

Table 1.20. *Values of Arrhenius parameters for the diffusion of various ions in silicate glasses*

Composition (mol%)	Species	Temp. range (°C)	D_0 (cm^2/s)	E_D (Kcal/mol)	Ref.
25K$_2$O–75SiO$_2$	^{22}Na	250 ~ 450	0.109	25.5	[111]
	^{42}K	250 ~ 450	3.71 × 10^{-4}	15.7	[111]
15Na$_2$O–10K$_2$O–75SiO$_2$	^{22}Na	250 ~ 450	0.13	24.3	[111]
	^{42}K	250 ~ 450	2.43	30.4	[111]
25Na$_2$O–75SiO$_2$	^{22}Na	250 ~ 450	8.19 × 10^{-4}	15.4	[111]
	^{42}K	250 ~ 450	0.158	27.6	[111]
20.0Na$_2$O–10.0CaO–70.0SiO$_2$	Na$^+$	250 ~ 450	5.37 × 10^{-3}	20.5	[112]
15.5Na$_2$O–12.8CaO–71.7SiO$_2$	Cs$^+$	450 ~ 535	3 × 10^{-8}	20.7	[113]
14.0Na$_2$O–12.0CaO–74.0SiO$_2$	^{42}K	347 ~ 565	1.25 × 10^{-2}	30.1	[114]
18.5Na$_2$O–7.4B$_2$O$_3$–74.1SiO$_2$	^{22}Na	381 ~ 565	3.5 × 10^{-3}	20.6	[115]
18.5Na$_2$O–7.4B$_2$O$_3$–74.1SiO$_2$	^{42}K	381 ~ 565	5.4 × 10^{-2}	29.2	[115]
29.8Na$_2$O–10.45Al$_2$O$_3$–59.70SiO$_2$	Na$^+$	354 ~ 490	2.0 × 10^{-3}	16.0	[116]
29.85Na$_2$O–10.45Al$_2$O$_3$–59.70SiO$_2$	Ag$^+$	354 ~ 490	2.2 × 10^{-3}	18.2	[117]
29.85Na$_2$O–10.45Al$_2$O$_3$–59.70SiO$_2$	K$^+$	354 ~ 473	6.8 × 10^{-3}	23.1	[116]

calculating the diffusion constant D from the values in Table 1.20 using Eq. (1.20), the values of D are in the range of $10^{-7} \sim 10^{-11}$ cm^2 s^{-1} in the vicinity of the glass transition temperature of ordinary silicate glasses. These values of $10^{-7} \sim 10^{-11}$ cm^2 s^{-1} are not very large and sometimes it takes a few months to complete the process for GRIN fabrication on a rod that is only a few millimeters in diameter.

Equation (1.20) shows that the diffusion constant of an alkali ion increases exponentially with temperature and so the ion-exchange can be enhanced by raising the treatment temperature. However, the suitable temperature range of ion exchange is not always the same and may differ depending on the application. For example, ion exchange for chemical tempering must be carried out at a temperature below the strain point of the glass, e.g. about 350 °C, to hinder the relaxation of the developed stress, whereas ion exchange for the formation of a gradient should be made at a temperature in the vicinity of the glass transition temperature so that the stress developed by exchange is relaxed.

Ion exchange has long been used in various applications without paying attention to the possible structural evolution of a glass. A recent study has shown that the occurrence of structural alteration of the glass which accompanies the remarkable change of glass properties depends on the ions to be incorporated [118]. Figure 1.25 shows an example of this structural alteration

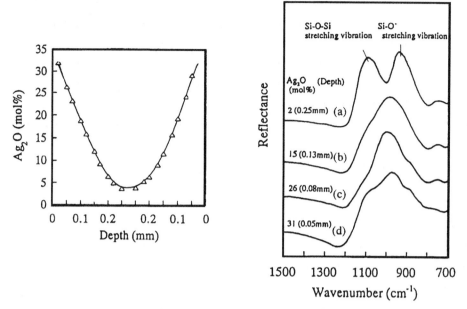

Fig. 1.25. Change of IR spectra with Na⁺ $\rightleftharpoons$ Ag⁺ ion-exchange. [Reprinted from M. Yamane, S. Shibata, A. Yasumori, T. Tano and H. Takada, *J. Non-Cryst Solids*, **203** (1996) 268, copyright (1996) with permission from Elsevier Science.]

observed by the Na⁺ $\rightleftharpoons$ Ag⁺ exchange in a sodium silicate glass of composition $33Na_2O-67SiO_2$ (mol%). The reflected infrared spectrum varies in a complicated manner due to the exchange of Ag⁺ for Na⁺. The structural evolution by the total replacement of Na⁺ by Ag⁺ led to a large reduction of T_g by nearly 200 °C, i.e. from the 460 °C of the original glass to 250 °C and the thermal expansion coefficient decreased from 12 ppm °C⁻¹ to 6 ppm °C⁻¹.

The electrical conductivity increased by about four orders of magnitude by the exchange of Na⁺ ions in a glass of composition $20Na_2O-10Al_2O_3-70SiO_2$ (mol%) with Cu⁺ ions using a fused salt bath consisting of a mixture of CuBr and CuI under a nitrogen atmosphere [107]. These results suggest further possibility for the application of the ion-exchange technique to the production of new functional materials with special properties.

References

1 *ASTM Standards on Glass and Glass Products*, prepared by ASTM Committee C14, American Society for Testing Materials, April 1955.
2 W. H. Zachariasen: *J. Am. Ceram. Soc.* **54** (1932) 3841.
3 B. E. Warren: *J. Appl. Phys.* **13** (1942) 602.

4 M. J. Weber: In *Handbook on the Physics and Chemistry of Rare Earth*, eds. K. A. Gschneider, Jr. and L. Eyring, (North-Holland, Amsterdam, 1979) p. 275.

5 M. B. Volf: *Mathematical Approach to Glass*, (Elsevier, Amsterdam, 1988) p. 21.

6a M. E. Nordberg, E. L. Mochel, H. M. Galfinkel and J. S. Olcott: *J. Am. Ceram. Soc.* **47** (1964) 215.

6b H. Kita, I. Kitano, T. Uchida, and M. Furukawa: *J. Am. Ceram. Soc.* **54** (1971) 321.

7 E. B. Shand: *Glass Engineering Handbook*, 2nd edn., (McGraw Hill, New York, 1958) p. 4.

8 Z. Strunad: *Glass-Ceramic Materials*, (Elsevier, Amsterdam, 1986) p. 9.

9 D. R. Uhlmann and H. Yinnon: in *Glass Science and Technology* Vol. 1, eds. D. R. Uhlmann and N. J. Kreidle, (Academic Press, New York, 1983) p. 1.

10 K. Matsusita and S. Sakka: *Bull Inst. Chem. Res., Kyoto Univ.* **59** (1981) 159.

11 G. W. Scherer: *Relaxation in Glass and Composites*, (Wiley, Chichester, 1986) p. 113.

12 H. R. Lillie: *J. Am. Ceram. Soc.* **16** (1933) 619.

13a M. Drexhage: in *Treatise on Materials Science and Technology*, Vol. 26, eds. M. Tomozawa and H. Doremus, (Academic Press, Orlando, 1985) p. 151.

13b J. A. Savage and S. Nielsen: *Infrared Phys.* **5** (1965) 197.

14 F. Ulbach: *Phys. Rev.* **92** (1953) 1324.

15 B. Bendow: *Appl. Phys. Lett.* **23** (1973) 133.

16 R. Olshansky: *Rev. Modern Phys.* **51** (1979) 341.

17a M. Nakahara: *Telecomm. J.* **48** (1981) 643.

17b T. Miya, T. Terunuma, T. Hosaka and T. Miyashita: *Electron. Lett.* **15** (1979) 106.

18 P. C. Schultz: *J. Am. Ceram. Soc.* **57** (1974) 309.

19 P. Kaiser, A. R. Tynes, H. W. Astle, A. D. Peason, W. G. French, R. E. Jeager and A. H. Cherin: *J. Opt. Soc. Am.* **63** (1973) 1141.

20 I. Fanderlik: *Optical Properties of Glass*, (Elsevier, Amsterdam, 1983) p. 78.

21 M. Ohno: in *Glass Handbook* (in Japanese), S. Sakka, T. Sakaino and K. Takahashi, ed., (Asakura Shoten, Tokyo, 1975) p. 71.

22a A. Yariv and R. A. Fisher: Chap. 1 in *Optical Phase Conjugation*, ed. R. A. Fisher (Academic Press, Inc., San Diego, 1983) p. 1.

22b P. N. Butcher and D. Cotter: Chap. 1 in *The Elements of Nonlinear Optics*, (Cambridge University Press, New York, 1990) p. 1.

22c Y. R. Shen: Chap. 4 in *The Principles of Nonlinear Optics* (John Wiley and Sons, Inc., New York, 1984) p. 53.

23 M. Cable: in *Glass Science and Technology*, Vol. 2, ed. D. R. Uhlmann and N. J. Kreidle, (Academic Press, Orlando, 1984) p. 1.

24 M. Wada: in *Glass Engineering Handbook*, ed. M. Yamane, I. Yasui, M. Wada, Y. Kokubu, K. Kondo and S. Ogawa, (Asakura Shoten, Tokyo, 1999) p. 7.

25a R. H. Dalton and M. E. Nordberg: US Patent 2 239 551, April 21 (1941).

25b J. F. Hyde: US Patent 2 272 342, February 10 (1942).

25c M. E. Nordberg: US Patent 2 326 059, August 3 (1943).

26 G. W. Scherer and P. C. Schultz: in *Glass Science and Technology*, Vol. 1, ed. D. R. Uhlmann and N. J. Kreidle, (Academic Press, New York, 1983) p. 49.

27 Data in *Handbook of Chemistry and Physics*, 66th edn. (1985–1986), (CRC Press, New York), Boca Raton.

28 K. Nassau and J. W. Shiever: *Am. Ceram. Soc. Bull.* **54** (1975) 1004.

29 P. C. Schultz: *J. Am Ceram. Soc.* **59** (1976) 214.

30 F. P. Kapron, D. B. Keck and R. D. Maurer: *App. Phys. Lett.* **17** (1970) 423.

31 D. B. Keck, P. C. Shultz and F. Zimar: US Patent 3 737 292, June 5 (1973).

32 J. B. MacChesney, P. B. O'Conner and H. M. Presby: *Proc. IEEE* **62** (1974) 1278.

33 T. Izawa, T. Miyashita and F. Hanawa: *Proc. Int. Conf. Integrated Optics and Optical Fiber Communication* (Tokyo, 1977) p. 375.

34 H. Dislich: *Glastech. Ber.* **44** (1971) 1.

35a G. W. Scherer and C. J. Brinker: *Sol-Gel Science* (Academic Press, Boston, 1990) p. 1.

35b M. Yamane: in *Sol-Gel Technology for Thin Film, Fiber, Preforms, Electronics, and Specialty Shapes*, ed. L. C. Klein, (Noyes, Park Ridge, 1988) p. 200.

35c G. W. Scherer and C. J. Luong: *J. Non-Cryst. Solids* **63** (1984) 16.

36 D. C. Bradley, R. C. Mehrotra and D. P. Gaur: *Metal Alkoxides*, (Academic Press, London, 1978) p. 42.

37 M. Yamane, S. Aso, S. Okano and T. Sakaino: *J. Mater. Sci.* **14** (1979) 607.

38a M. Yamane and S. Noda: *J. Ceram. Soc. Jpn.* **101** (1993) 11.

38b M. Yamane: in *Sol-Gel Optics; Processing and Applications*, ed. L. C. Klein, (Kluwer Academic, Boston, 1994) p. 391.

39 T. Kawaguchi, H. Hishikura, J. Iura and Y. Kokubu: *J. Non-Cryst. Solids* **63** (1984) 61.

40a S. Wallence and L. L. Hench: in *Better Ceramics Through Chemistry*, ed. C. J. Brinker, D. E. Clark and D. R. Ulrick, (North Holland, New York, 1984) p. 47.

40b T. Adachi and S. Sakka: *J. Non-Cryst. Solids* **100** (1988) 250.

41a C. A. M. Mulder, J. G. van Lierop and G. Frens: *J. Non-Cryst. Solids* **82** (1986) 92.

41b J. G. van Lierop, A. Huizing, W. C. P. M. Meerman and C. A. M. Mulder: *J. Non-Cryst. Solids* **82** (1986) 265.

42a E. M. Rabinovich, D. W. Johnson, J. B. MacChesney and E. M. Vogel: *J. Non-Cryst. Solids* **47** (1982) 435.

42b M. Toki, S. Miyashita, T. Takeuchi, S. Kanbe and A. Kochi: *J. Non-Cryst. Solids* **100** (1988) 479.

43 Y. Sano, S. H. Wang, R. Chaudhuri and A. Sarker: *SPIE* Vol. 1328, *Proc Sol-Gel Optics*, (San Diego, 1990) p. 52.

44 J. Livage, F. Babonneau and C. Sanchez: in *Sol-Gel Optics; Processing and Applications*, ed. L. C. Klein, (Kluwer Academic, Boston, 1994) p. 39.

45 B. E. Yoldas: *J. Non-Cryst. Solids* **38, 39** (1980) 81.

46 M. Yamane: *Bull. Inst. Chem. Res., Kyoto Univ.* **72** (1994) 254.

47 S. Sakka: *SPIE* Vol. 1758, *Proc. Sol-Gel Optics II*, (San Diego, 1992) 2.

48 I. Fanderlik: *Silica Glass and its Application*, (Elsevier, Amsterdam, 1991) p. 15.

49 G. H. Sigel, Jr.: *Treatise on Materials Science and Technology*, Vol. 12, ed. M. Tomozawa and H. Doremus, (Academic Press, New York, 1977) p. 5.

50 S. Liu and L. L. Hench: *SPIE*, 1758, *Proc. Sol-Gel Optics II*, San Diego (1992) 14.

51 N. J. Kridle: *Glass Science and Technology*, Vol. 1, eds. D. R. Uhlmann and N. J. Kreidle (Academic Press, New York, 1983) p. 107.

52 *Data Book of Glass Composition, 1991* (Glass Manufacturer's Association, Japan, 1991) pp. 76.

53a M. J. Weber: *J. Non-Cryst. Solids* **123** (1990) 208.

53b M. J. Weber, C. B. Layne and R. A. Saroyan: *Laser Program Annual Report - 1975* (Lawrence Livermore Laboratory, 1976) p. 192.

54 W. A. Weyl: *Colored Glasses*, Society of Glass Technology (1953).

55 E. F. Riebling: *J. Chem. Phys.* **55** (1971) 804.

56 E. N. Boulos and N. J. Kreidl: *J. Am. Ceram. Soc.* **54** (1971) 368.

57 J. L. Pinguet and J. E. Shelby: *J. Am. Ceram. Soc.* **68** (1985) 450.

58a R. F. Bartholomeu: *J. Non-Cryst. Solids* **7** (1972) 221.

58b M. Yamane, S. Shibata, T. Yano and K. Azegami: unpublished data.

59a T. Takamori: *Treatise on Materials Science and Technology*, Vol. 17, ed. M. Tomozawa and H. Doremus (Academic Press, New York, 1979) pp. 173.

59b Y. Asahara and T. Izumitani: US Pat. No. 3 885 974 (1975).

59c M. Ishiyama, T. Matsuda, S. Nagahara and Y. Suzuki: *Asahi Garasu Kenkyu Hokoku* **16** (1966) 77.

59d P. P. Pirooz: US Pat. No. 3 088 833 (1963).

59e J. W. Malmendier: US Pat. No. 3 883 358 (1975).

60a C. B. Rubinstein, S. B. Berger, L. G. Van Uitert and W. A. Bonner: *J. Appl. Phys.* **35** (1964) 2338.

60b K. Tanaka, K. Hirao and N. Soga: *Jpn. J. Appl. Phys.* **34** (1995) 4825.

61 M. E. Lines: *J. Non-Cryst. Solids* **103** (1988) 279.

62a M. E. Lines, J. B. MacChesney, K. B. Lyons, A. J. Bruce, A. E. Miller and K. Nassou: *J. Non-Cryst. Solids* **107** (1989) 251.

62b P. L. Higby and I. D. Aggarwal: *J. Non-Cryst. Solids* **163** (1993) 303.

62c S. Takahashi and I. Sugimoto: *J. Lightwave Technol.* **LT-2** (1984) 613.

63a R. Adair, L. L. Chase and S. A. Payne: *J. Opt. Soc. Am.* **B4** (1987) 875.

63b N. F. Borrelli, B. G. Aitken, M. A. Newhouse and D. W. Hall: *J. Appl. Phys.* **70** (1991) 2774.

63c H. Nasu, O. Matsushita, K. Kamiya, H. Kobayashi and K. Kubodera: *J. Non-Cryst. Solids* **124** (1990) 275.

63d S. H. Kim, T. Yoko and S. Sakka: *J. Am. Ceram. Soc.* **76** (1993) 2486.

64a J. D. Mackenzie: *J. Am. Ceram. Soc.* **43** (1960) 615.

64b M. Munakata: *Solid State Electronics* **1** (1960) 159.

65 M. Poulain, M. Poulain, J. Lucas and P. Brun: *Mat. Res. Bull.* **10** (1975) 243.

66 C. M. Baldwin, R. M. Almeida and J. D. Mackenzie: *J. Non-Cryst. Solids* **43** (1981) 309.

67 S. Shibata, M. Horiguchi, K. Jinguji, S. Mitachi, T. Kanamori and T. Manabe: *Electron. Lett.* **17** (1981) 775.

68a P. W. France, M. G. Drexhage, J. M. Parker, M. W. Moore, S. F. Carter, and J. V. Wright: *Fluoride Glass Optical Fiber*, (CRC Press, Boca Raton, 1990).

68b S. F. Carter, M. W. Moore, D. Szabesta, J. Williams, D. Ranson and P. W. France: *Electron. Lett.* **26** (1990) 2115.

69a P. A. Tick: *Mater. Sci. Forum.* **32–33** (1988) 115.

69b T. Kanamori: *Mater. Sci. Forum.* **19–20** (1987) 363.

70 J. Coon and J. E. Shelby: *Mater. Sci. Forum.* **32–33** (1988) 37.

71 S. Mitachi and P. A. Tick: *Mater. Sci. Forum.* **32–33** (1988) 197.

72 P. W. France, M. G. Drexhage, J. M. Parker, M. W. Moore, S. F. Carter and J. V. Wright: *Fluoride Glass Optical Fibers*, (CRC Press, Boca Raton, 1990).

73 P. A. Tick: *Mater. Sci. Forum.* **5** (1985) 165.

74 T. Yano, J. Mizuno, S. Shibata, M. Yamane, S. Inoue and Y. Onoda: *J. Non-Cryst. Solids* **213, 214** (1997) 345.

75 M. J. Weber, C. B. Layne and R. A. Saroyan: *Laser Program Annual Report – 1975*, (Lawrence Livermore Laboratory, 1976) p. 192.

76 T. Minami: *J. Non-Cryst. Solids* **56** (1983) 15.

77a L. M. Sanford and P. A. Tick: US Pat. No. 4 314 031 (1982).

77b P. A. Tick: *Phys. Chem. Glasses* **25** (1984) 149.

78 M. Anma, T. Yano, A. Yasumori, H. Kawazoe, M. Yamane, H. Yamanaka and M Katada: *J. Non-Cryst. Solids* **135** (1991) 79.

79 S. Tsuchihashi, A. Yano, T. Komatsu and K. Adachi: *Yogyo-Kyokai-Shi*, **74** (1966) 353.

80 A. E. Bell and R. A. Bartolini: *Appl. Phys. Lett.* **34** (1979) 275.

81a N. F. Mott and E. A. Davis: *Electronic Processes in Non-Crystalline Materials*, (Oxford University Press, Oxford, 1979) p. 442.

81b A. D. Pearson, W. R. Northover, J. F. Dewald and W. F. Peck, Jr.: *Advances in Glass Technology* **1** (1962) 357.

81c R. Hilton and M. Brau: *Infrared Phys.* **3** (1963) 69.

81d J. A. Savage and S. Nielsen: *Phys. Chem. Glasses* **5** (1964) 82.

81f J. T. Edmond: *J. Non-Cryst. Solids* **1** (1968) 39.

82 S. G. Bishop and D. L. Mitchell: *Phys. Rev.* **B8** (1973) 5696.

83 E. A. Fagen and H. Fritzsche: *J. Non-Cryst. Solids* **2** (1970) 180.

84 R. A. Street: *Solid State Commun.* **24** (1977) 363.

85 S. R. Ovshinsky: *Phys. Rev. Lett.* **21** (1968) 1450.

86a A. G. Pincas: *Application in Nucleation and Crystallization in Glasses*, (American Ceramic Society, 1971) p. 210.

86b G. H. Beall and D. A. Duke: *J. Mater. Sci.* **4** (1969) 340.

86c J. Nagasaki: *Verres Refr.* **20** (1966) 477.

86d Corning Glass Works: *Properties of Glasses and Glass Ceramics* (1972) Corning.

87 Corning Glass Works: *Pyroceram*, Code 9606.

88 Corning Glass Works: *Macor, Machinable Glass Ceramics*, Code 9658.

89 T. Kokubo, M. Shigematsu, R. Nagashima, M. Tashiro, T. Nakamura, T. Yamamuro and S. Higashi: *Bull. Inst. Chem. Res., Kyoto Univ.* **60** (1982) 260.

90 Nippon Electric Glass Co.: *Process and Product* (1981) 10.

91 X. Zhou and M. Yamane: *J. Ceram. Soc. Jpn.* **96** (1988) 152.

92 P. F. James: *Nucleation and Crystallization in Glasses*, ed. J. Simmons, D. R. Uhlmann and G. H. Beall, (American Ceramic Society, Columbus, 1982).

93a W. A. Weyl: *Colored Glasses*, (Society of Glass Technology, 1951) p. 329.

93b C. R. Bamford: *Color Generation in Glasses*, (Elsevier, Amsterdam, 1977).

93c A. E. Badger, W. A. Weyl and H. Rudow: *Glass Ind.* **20** (1939) 407.

94 J. W. Haus, N. Kalayaniwalla, R. Inguva, M. Bloemer and C. M. Bowden: *J. Opt. Soc. Am.* **B6** (1989) 797.

95 R. J. Araujo: *Treatise on Materials Science and Technology*, Vol. 12, eds. M. Tomozawa and R. H. Doremus (Academic Press, New York, 1977) p. 91.

96a V. N. Filipovich: *Phase Separation in Glass*, eds. O. V. Mazurin and E. A. Porai-Koshits, (North Holland, Amsterdam, 1984) p. 15.

96b O. V. Mazurin, G. P. Roskova and E. A. Porai-Koshits: *ibid*, p. 103.

96c M. Tomozawa: *Treatise on Materials Science and Technology*, Vol. 17, eds. M. Tomozawa and R. H. Doremus, (Academic Press, New York, 1979) p. 71.

97 M. Yamane: unpublished data.

98 H. P. Hood and M. E. Nordberg: U.S. Patent 2 221 709 (1940).

99 J. J. Hammel: U.S. Patent 3 843 341 (1974).

100 T. Nakajima and M. Shimizu: Ceramics. *Bull. Ceram. Soc. Jpn* **21** (1986) 408.

101 T. Kokubu and M. Yamane: *J. Mater. Sci.* **20** (1985) 4309.

102 T. H. Elmer and H. E. Meissner: *J. Am. Ceram. Soc.* **58** (1976) 206.

103 S. Omi, H. Sakai, Y. Asahara, S. Nakayama, Y. Yoneda and T. Izumitani: *Appl. Opt.* **27** (1988) 797.

104 R. F. Bartholomew and H. M. Garfinkel: *Glass Science and Technology*, Vol. 5,

eds. D. R. Uhlmann and N. J. Kreidl, (Academic Press, New York, 1980) p. 217.

105 M. E. Nordberg, E. L. Mochel, H. M. Garfinkel and J. S. Olcott: *J. Am. Ceram. Soc.* **47** (1964) 215.

106 H. Kita, I. Kitano, T. Uchida and M. Furukawa: *J. Am. Ceram. Soc.* **54** (1971) 321.

107 J. Lee, T. Yano, S. Shibata and M. Yamane: *J. Non-Cryst. Solids* **222** (1997) 120.

108 M. Oikawa and K. Iga: *Appl. Opt.* **21** (1982) 1052.

109 R. H. Doremus: *J. Phys. Chem.* **68** (1964) 2212.

110 N. P. Bansal and R. H. Doremus: *Handbook of Glass Properties*, (Academic Press, New York, 1986) p. 500.

111 J. W. Fleming and D. E. Day: *J. Am. Ceram. Soc.* **55** (1972) 186.

112 R. Terai: *Phys. Chem. Glasses* **10** (1969) 146.

113 L. Kahl and E. Schiewer: *Atom wirtschaft* **16** (1971) 434.

114 E. Richter, A. Kolitsch and M. Hahnert: *Glastech. Ber.* **55** (1982) 171.

115 M. Hahnert, A. Kolitsch and E. Richter: *Z. Chem.* **24** (1984) 229.

116 A. Kolitsch, E. Richter and W. Hinz: *Silikattechnik* **31** (1980) 41.

117 A. Kolitsch and E. Richter: *Z. Chem.* **21** (1981) 376.

118 M. Yamane, S. Shibata, A. Yasumori, T. Yano and H. Takada: *J. Non-Cryst. Solids* **203** (1996) 268.

2

Gradient index glass

Introduction

In a conventional optical system, the refractive index within each optical component is considered to be homogeneous. In the design of such systems, the curvatures, thickness, and refractive index of each component are varied independently to optimize the performance of a lens system. It is possible, on the other hand, to manufacture lens elements whose index of refraction varies in a continuous way as a function of spatial coordinate. Such materials, which are often said to be GRadient INdex (GRIN) materials, have various advantages for use in both focusing and imaging purposes.

There are three types of refractive index gradients depending upon the type of symmetry [1–3]. The first is the axial gradient (a-GRIN), in which the refractive index varies in a continuous way along the optical axis. These gradients are particularly useful for replacing aspheric surfaces in monochromatic systems, e.g. in collimators for laser beams.

The second type of gradient is the radial gradient (r-GRIN). The refractive index n in this case varies perpendicular to and continuously outward from the optical axis. This type of gradient index lens, often called a GRIN rod lens when the diameter is small, exhibits the property that light propagating parallel to the optical axis of symmetry can be focused periodically if n varies as an appropriate function of its radial distance from the axis of symmetry. The main applications of gradient index materials in use today are based only on this type of gradient. In this type the problems of manufacturing large geometry lenses has been the limiting feature in implementing gradient index optics in photographic systems for many years. There are now numerous techniques which enable the fabrication of many different r-GRIN glass materials.

The third type is the spherical gradient (s-GRIN), where the refractive index n is symmetric about a point. Although no optical example of the full spherical gradient has ever been developed, a two-dimensional array of microlenses with a hemispherical gradient index have been developed for various attractive applications.

This chapter reviews the major current application of r-GRIN glass materials, the design of gradient index profiles, and the processes and parameters for fabricating them.

2.1 Applications

The applications of r-GRIN materials are divided into two major areas. The first applications are for focusing purposes. In this case, GRIN rod lenses of a small diameter and flat surface are currently used for optical telecommunications, signal processing [4–6], and micro-optical devices such as an optical disk [7–9]. If the gradient index profile is chosen properly, the propagation path of the light ray varies sinusoidally along the rod axis. Using such lenses, collimated light can be focused or diverging light can be collimated, like a convex lens. Such lenses are therefore useful for focusing or collimating light emitting from a narrow-diameter source such as an optical fiber and a laser diode, for optically coupling a light source with an optical fiber and for coupling two fibers, as shown in Fig. 2.1. Various designs for such GRIN rod lens devices, including connectors, attenuators, directional couplers, switches, isolators and wavelength-division multiplexers, have been proposed [10].

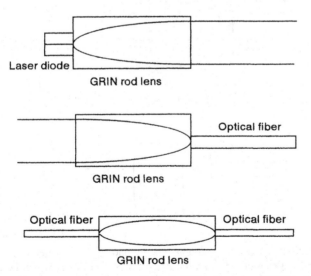

Fig. 2.1. Schematic cross section of a coupler and collimator using GRIN rod lenses with small diameter and flat surface.

The second major applications are for imaging purposes. These applications are subdivided into two sections. One is a most important and commercially successful application of small-diameter GRIN lenses in copy machines [11] and facsimiles [12]. The single imaging lens in copy machines is replaced by a two- or three-row array of GRIN rod lenses 1 cm long and 1 mm in diameter. This array forms an image in a similar manner to insect eyes by superimposing parts of the image through each of the rod lenses. An assemblage of 600 to 2000 rod lenses can obtain a field of view as wide as a 36-inch-wide sheet of paper. The most important feature of the GRIN lens array is that the distance from the original surface to the image surface, namely the conjugate length, is short. When a conventional lens is used in a copy machine, its conjugate length becomes as long as 600 ∼ 1200 mm, whereas in the case of a GRIN lens array the conjugate length is less than 60 mm and a very compact imaging system can be obtained. Recently copy machines used for office work have been required to be capable of magnifying operation, and so the use of GRIN lens arrays is now limited to compact copy machines.

The r-GRIN material is also useful as a part of the camera lens and the photographic objective lens [13]. The number of lenses can be reduced by a factor of two and enormous weight and cost savings can thereby be achieved if a sufficiently large-diameter r-GRIN material can be fabricated. As shown in Fig. 2.2 for instance, the new double Gauss photographic objective consisting

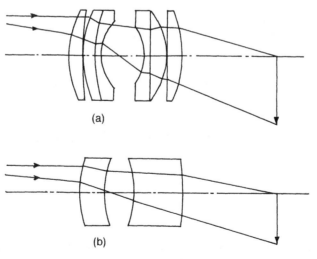

Fig. 2.2. Configurations of (a) new double Gauss photographic objective, and (b) GRIN photographic objective. [Reprinted from L. G. Atkinson, S. N. Houde-Walter, D. T. Moore, D. P. Ryan and J. M. Stagaman, *Appl. Opt.* **21** (1982) 993, copyright (1982) with permission from the Optical Society of America.]

of six elements can be configurated by the combination of only two r-GRIN lenses and the third-order spherical aberration, coma, astigmatism and Petzval field curvature are all corrected to less than 1 μm [13]. By suitable selection of design parameters, chromatic aberration can also be corrected [13]. Another example is the application of GRIN lens to a zoom lens, which provides a reduction in the total lens length of the system by 40% [14].

2.2 Design of radial gradient index profiles

When light enters a GRIN material in which n decreases along the radial distance from the central axis, it will be bent toward the axis. The property of special interest is that all rays starting at a given object point at one end converge to a single image point at the other end, that is, all rays propagate with the same period (pitch) of $p = 2\pi/g$. Thus parallel rays focus at a point as if they had passed through a conventional lens, and alternately and periodically converge and diverge. For meridional rays (rays lying in a plane containing the lens axis), the ideal index profile satisfying the conditions described above is represented in the form of a hyperbolic secant as follows [15, 16]:

$$n^2(r) = n_0^2 \operatorname{sech}^2(gr)$$
$$= n_0^2[1 - (gr)^2 + h_4(gr)^4 - h_6(gr)^6 + \ldots] \tag{2.1}$$

when n_0 is the refractive index at the center axis, r is the radial distance from the rod center axis, g is a focusing constant determining the power of the lens, and $h_4(= 2/3)$ and $h_6(= 17/45)$ are higher order dimensionless parameters. For helical rays (rays that spiral down the lens at a constant radius) the ideal distribution is of the form

$$n^2(r) = n_0^2[1 + (gr)^2]^{-1}$$
$$= n_0^2[1 - (gr)^2 + (gr)^4 - (gr)^6 + \ldots]. \tag{2.2}$$

It is clear that there is no gradient index profile that can provide perfect imaging for all rays.

In the paraxial approximation and if small index changes are considered, the index distribution can be closely approximated by a parabola

$$n^2(r) = n_0^2(1 - (gr)^2) \tag{2.3}$$

or [17]

$$n(r) = n_0\{1 - (g^2/2)r^2\} \tag{2.4}$$

or [13, 18]

$$n(r) = N_{00} + N_{10}r^2 + N_{20}r^4 + \ldots \tag{2.5}$$

By analogy with lens optics the term $n_0 g r_0$ is called the numerical aperture (NA) of the r-GRIN lens.

2.3 Parameters for fabrication of index gradient

2.3.1 Gradient index profile

The most important parameter of the GRIN materials is the gradient index profile. According to the Lorentz–Lorenz formula, the refractive index of a substance is related to the number of constituent ions (N_i) and to their electronic polarizabilities (α_i) by the following equation [19]:

$$(n^2 - 1)(n^2 + 2) = (4\pi/3) \sum N_i \alpha_i. \tag{2.6}$$

A given gradient index profile can be developed therefore, by the precise control of the concentration profile of the constituent ions (index-modifying ions). The shape of the concentration profiles can be controled by the simple diffusion process of the index-modifying ions.

From elementary diffusion calculations [20], if the concentration of interested ions $C(r, t)$ is initially uniform C_1 throughout the rod radius r_0 and C_0 at the surface, the boundary conditions are

$$C(r, t) = C_0, \; r = r_0, \; t > 0 \tag{2.7}$$

$$C(r, t) = C_1, \; 0 < r < r_0, \; t = 0 \tag{2.8}$$

and the solution of the diffusion equation for a rod of finite radius ($r < r_0$) is then expressed as the following equations,

$$\{C(x, t) - C_0\}/(C_1 - C_0) = 2 \sum_{n=1}^{\infty} \exp(-\beta_n^2 T_f)[J_0(\beta_n, x)/\beta_n J_1(\beta_n)] \tag{2.9}$$

where T_f is the diffusion parameter ($= Dt/r_0^2$), β_n are roots of $J_0(\beta_n) = 0$, x is the radial parameter ($= r/r_0$), J_0 is the Bessel function of the first kind of order zero, D is the diffusion coefficient, t is the treatment time, J_1 is the Bessel function of first order and r_0 is the rod radius. Figure 2.3 demonstrates how variations in T_f affect the concentration profiles for a r-GRIN material along its radial distance. The concentration profile still remains flat near the center at $T_f < 0.05$, but seems to be parabolic at $T_f > 0.05$.

If the refractive index depends linearly on concentration, then a given concentration profile can be translated into that of a refractive index profile, $n(r, t)$. At the appropriate diffusion parameter, T_0 (larger than 0.05), Eq. (2.9) can be expressed as a power series in the radial parameter x ($= r/r_0$) [17],

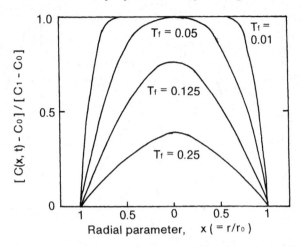

Fig. 2.3. Variations of concentration profile with variations in T_f.

$$\{C(x,\ t) - C_0\}/(C_1 - C_0) = 2 \sum_{n=1}^{\infty} \exp(-\beta_n^2 T_0)\{1/\beta_n J_1(\beta_n)\}y \qquad (2.10)$$

$$y = (1/1!)(\beta_n/2)^2 (r/r_0)^2 - (1/2!)^2 (\beta_n/2)^4 (r/r_0)^4$$
$$+ (1/3!)(\beta_n/2)^6 (r/r_0)^6 - \dots . \qquad (2.11)$$

Then, if the secondary differential of Eq. (2.11) is zero,

$$d^2 y/dr^2 = (\beta_n/2)^2 [2 - 3(\beta_n/2)^2 (r/r_0)^2 + (5/6)(\beta_n/2)^4 (r/r_0)^4 - \dots = 0$$
$$(2.12)$$

and $(r/r_0) = 0.82$. Thus, the inflection point in the concentration curve, i.e. in Eq. (2.11), appears at $r = 0.82\ r_0$, suggesting that an approximately parabolic refractive index distribution over 80% of the diameter of the r-GRIN materials will be produced by the diffusion process [21].

Since the primary interest is a radial parabolic index profile given by Eq. (2.3), the best fit may also be achieved by minimizing the standard deviation, defined as [22]

$$dS = \left[\sum_i \{n(x_i,\ t)^2 - n_0(1 - g^2 x_i^2)\}^2 / N \right]^{1/2} . \qquad (2.13)$$

Figure 2.4 shows the variations of the focusing constant g and the standard deviation dS as a function of the diffusion parameter T_f when the least-squares parabolic fit of Eq. (2.9) within 60% radius from the center axis was carried out under the condition that the maximum index difference between the center and the periphery, Δn is 0.03 [23, 24]. The results indicate that the best fit parabolic case is obtained at $T_f = 0.1$.

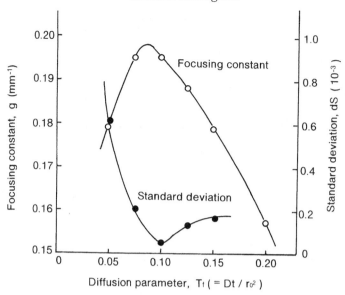

Fig. 2.4. Focusing constant g and standard deviation dS as a function of diffusion parameter T_f ($\Delta n = 0.03$, $r_0 = 1$ mm). [Reprinted from Y. Asahara and S. Ohmi, A digest of the technical papers, the forth topical meeting on gradient-index optical imaging systems, Monterey (1984) ThE-B1, coypright (1984) with permission from the Optical Society of America.]

2.3.2 Maximum index difference

The second parameter that should be considered is the maximum index difference, Δn, between the center and the periphery of the GRIN rod. This parameter is closely related to the refractive index at the center axis of the rod n_0, the focusing constant g and thus the numerical aperture of the r-GRIN lens $n_0 g r_0$. If an index-modifying ion is very polarizable and diffuses with ease in glass above the annealing temperature, the concentration gradient of the ion resulting from diffusion will cause a large variation of the refractive index. The ionic radii [25], electronic polarizabilities [25] and refractive constant [26] of index-modifying ions are listed in Table 2.1. Thallium, cesium, silver and in some cases, lead, germanium and titanium are selected as refractive index-modifying ions because of their degree of polarizability and mobility in glass.

2.3.3 Chromatic property

The chromatic aberration is another important parameter in addition to the precision of the refractive index profile, geometry and Δn, that must be taken

Table 2.1. *Ionic radii and electronic polarizabilities of index-modifying ions* [25, 26]

Ion	Ionic radius [25] (Å)	Electronic Polarizability [25] (Å³)	Refractive constant [26] a_d	Dispersion constant [26] a_{F-C}
Li^+	0.78	0.03	4.60	0.10
Na^+	0.95	0.41	6.02	0.16
K^+	1.33	1.33	9.54	0.20
Rb^+	1.49	1.98	12.44	0.24
Cs^+	1.65	3.34	17.48	0.31
Tl^+	1.49	5.20	31.43	2.55
Ag^+	1.26	2.4	15.97	0.63
Mg^{2+}	0.78	0.09	–	–
Ca^{2+}	0.99	1.1	12.66	0.26
Sr^{2+}	1.27	1.6	16.50	0.33
Ba^{2+}	1.43	2.5	19.80	0.34
Zn^{2+}	0.83	0.8	12.20	0.34
Cd^{2+}	1.03	1.8	–	–
Pb^{2+}	1.32	4.9	28.41	1.30
Si^{4+}	0.41	0.017	12.51	0.19
Ti^{4+}	0.68	0.19	25.00	1.34
Ge^{4+}	0.53	–	17.47	0.42

into consideration in the fabrication of r-GRIN lens elements for imaging applications such as cameras, binoculars and color copiers. The dispersion $n_F - n_C$ and Abbe number, which represent chromatic properties, usually vary along the radius in the GRIN materials, based on the concentration gradient of the index-modifying ions.

The total paraxial axial color (*PAC*) of a gradient index lens is given by [13]

$$PAC = Y(\Delta\Phi_{surf} + \Delta\Phi_{grad}) \qquad (2.14)$$

where Y is a constant related with the height of the axial ray and its final angle. $\Delta\Phi_{surf}$ and $\Delta\Phi_{grad}$ are the difference in power of the surfaces and in the power of the GRIN profile respectively, due to each wavelength-dependence of the base index N_{00} and of the coefficient N_{10} in Eq. (2.5) and are given by

$$\Delta\Phi_{surf} = \Phi_{surf}/V \qquad (2.15)$$

$$\Delta\Phi_{grad} = \Phi_{grad}/V_{10}. \qquad (2.16)$$

Here Φ_{surf} and Φ_{grad} are the power due to the surface and the gradient respectively, V is the Abbe number of the base index N_{00}

$$V = (N_{00d} - 1)/(N_{00F} - N_{00C}) \qquad (2.17)$$

and V_{10} is the dispersion for coefficient N_{10}

$$V_{10} = N_{10d}/(N_{10F} - N_{10C}). \tag{2.18}$$

The total *PAC* for a radial gradient lens singlet will be eliminated if

$$\Phi_{\text{surf}}/V = -\Phi_{\text{grad}}/V_{10}. \tag{2.19}$$

The dispersion change of some univalent and divalent ion-exchange pairs can be estimated from the parameter a_{F-C} in Table 2.1 given by S. D. Fantone [26]. The chromatic aberration of r-GRIN lenses can be minimized by selecting a suitable ion-exchange pair for a given base glass, just in the same way that the chromatic aberration of the achromatic lens is canceled by the combination of concave lens (high index, low dispersion) and convex lens (low index, high dispersion) [27–30].

2.3.4 Stress

The concentration gradient of index-modifying ions produces not only a change of optical properties, but also a change of other material properties with respect to the variation of the glass volume. One of the most important properties caused by the concentration gradient is the stress, which builds in a birefringence in GRIN materials via a photoelastic effect. In the simple interdiffusion process, the reduction in the amount of index-modifying ions is compensated by the counter-diffusing ions. Therefore, there is less difference in density between the center and the periphery. Two factors may be responsible for the stress: the ionic radius mismatch and the thermal expansion mismatch. When the interdiffusion temperature is much lower than the stress relaxation temperature or when the viscosity of glass is very high, for instance, in an alkali-free high-silica glass, the surface cannot expand freely and the ionic radius mismatch results in compressive or tensile stress in the periphery. In most cases, since the interdiffusion temperature is near the stress relaxation temperature, the glass accommodates the interdiffusion of two ions and no such discrepancy was observed. Thus the permanent stress which develops in r-GRIN materials is attributed to the change in the thermal expansion coefficient from the center toward the periphery caused by the thermal expansion mismatch. When the sample is cooled down from the interdiffusion temperature, this effect will generate stress. With the exception of any special case, the stress is usually compressive or much smaller than the strength of the materials, so that it can be ignored.

2.4 Materials and fabrication techniques

2.4.1 Ion exchange

2.4.1.1 Conventional ion-exchange technique

The most common technique for fabricating GRIN materials is by an ion-exchange process [17] as shown in Fig. 2.5. This is probably the most widely used technique because it is the simplest in terms of instrumentation and control. In the simplest configuration, a glass containing a highly polarizable and monovalent ion such as cesium or thallium is immersed in a molten salt bath containing another low polarizable ion such as potassium or sodium. In this process, ions from the salt bath diffuse into a glass and exchange for index-modifying ions in the glass and therefore a gradient in composition is created. The temperature of the treatment is chosen within the range to increase the interdiffusion rate and to relax the stress in the glass that is induced by the ion size mismatch. The refractive index profile is mainly determined by the distribution of index-modifying ion in the glass rod. Lenses with a nearly parabolic index profile were fabricated using this technique on thallium-containing borosilicate glasses [17].

A gradient index lens with high numerical aperture and also with desirable higher order parameters has been fabricated by a new modified ion-exchange technique [31, 32]. In this technique, a glass rod containing a highly polarizable ion is immersed in a KNO_3 salt bath containing an ion with a medium

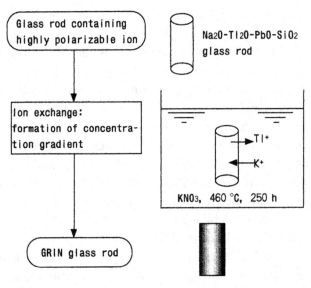

Fig. 2.5.   Ion exchange process for fabricating r-GRIN rod.

electronic polarizability. The medium polarizable ion penetrates only into the marginal region of the rod and modifies the inappropriate index gradient in this region. The GRIN lens fabricated by the modified technique has a high numerical aperture and a low sperical aberration.

These lenses, however, suffer from severe chromatic aberration in optical image transmission under white light. The dispersion at the periphery is lower than that at the center because the ion with high electronic polarizability was decreased by ion exchange. For instance, the Abbe number of a GRIN rod lens based on thallium-lead silicate is calculated to be 37.6 at the center and 43.0 at the periphery of the rod [17].

The chromatic aberration of these lenses can be minimized by selecting a suitable ion-exchange pair for a given base glass. As is clear in Table 2.1, one example is a K^+–Cs^+ exchange, which suffers much less chromatic aberration than a lens with conventional K^+–Tl^+ exchange and similarly produced achromatic GRIN lenses [33]. Another example is the fabrication based on ion pairs of Li^+ and Na^+ [34, 35].

Lenses of $1 \sim 4$ mm diameter, $g = {\sim}0.430$ mm^{-1}, $n_0 = {\sim}1.66$ and NA $= {\sim}0.60$ have been made by the ion-exchange technique and are now successfully commercialized [36]. These lenses find unique applications in the field of micro-optics devices such as lens arrays for copy machines, relay rods for endoscopes, pickup and recording lenses for compact discs, and connectors for optical communications.

2.4.1.2 Modified ion-exchange technique

The ion-exchange technique has limitations for large geometry lenses, which creates severe problems for radial gradients in photographic applications. The maximum diameter of rod lenses produced by this technique has been only a few millimeters or even less. The difficulty is due to the low diffusion coefficient of the index-modifying ions, although this depends also on the temperature and base glass. Various attempts have been made to reduce the fabrication time in ion-exchange processes.

One possible method is to carry out the ion exchange at a temperature above the glass transition point. The quick formation of an index profile is possible by the increase of mutual diffusion coefficients of ions by two to three orders of magnitude compared with the exchange at a temperature below T_g. Lenses of 20 mm in diameter have been obtained by ion exchange at well above the glass transition temperature T_g for 18 days [37].

Ion exchange of Ag^+–Na^+ is thought to be of possible use for making GRIN rod lenses of large diameter because of their high interdiffusion coefficients in the glass. Although the ionic radius of Ag^+ is smaller than that of K^+, the ionic radius mismatch is less pronounced than that of the Na^+–K^+ pair. In most cases, the stress is compressive and much smaller than the strength. Ag^+–Na^+ exchange yields satisfactory values of Δn and the stress effects are negligible compared with the polarizability change. Lenses of about 4 mm in diameter with Δn of 0.04 have been prepared by exchanging Na^+ for Ag^+ at 480 °C for 72 h on an Ag_2O–Al_2O_3–P_2O_5 system [38]. Assuming the diameter required for GRIN lenses for cameras and microscopes to be 15 mm, the fabrication time is estimated to be about 1100 h, which is a great saving of fabrication time over K^+–Tl^+ ion exchange. Although the index distribution is concave, rod lens of 37 mm in diameter was obtained, using a cylindrical rod of alumino-silicate glass of 40 mm in diameter and 50 mm long and Ag^+–Na^+ exchange process only for 960 h [39].

The problem with ion exchange in a silver-containing system is that it is not possible to melt large amounts of Ag_2O into silicate glass. The Ag ions are reduced in the glass and tend to be colloidal, so that a sufficiently high concentration of the Ag ions cannot be readily introduced into the glass by conventional melting and cooling processes. Therefore, aluminophosphate glass is the only system that can be made by this technique [38, 40]. Another problem to be noted concerning this technique is that the surface of the glass rod is severely attacked by molten salt and the diameter of the eventual rod is remarkably reduced. For instance, an original diameter of 4.8 mm changed to 4.0 mm during the ion exchange [38].

A modification of the ion-exchange process for improving the problem of Ag reduction is a double ion-exchange technique [41, 42]. The process is composed of two steps. The starting material is a stable and optically homogeneous glass containing, for instance, about 40 mole% of Na_2O. The glass rod is initially immersed in a molten salt bath containing $AgNO_3$ and Ag ions are diffused into the glass rod until the concentration of Ag ions becomes uniform (ion stuffing). The rod is then soaked again in a salt bath containing $NaNO_3$ and the Ag ions in the glass structure are redissolved (ion unstuffing). The concentration gradient is controllable at this step by adjusting the treatment temperature and time, in the same way as in the ion-exchange process. The diffusion coefficient of Ag ions crudely calculated from the concentration profile is 6.5×10^{-8} cm^2/s. This value is much higher than the interdiffusion constant of a Tl^+–K^+ ion pair during the conventional ion exchange.

Although this technique requires a relatively long fabrication time due to a very long extra step for ion stuffing, this technique offers the possibility of making large diameter GRIN rod lenses. A rod lens of 9.7 mm diameter was fabricated by stuffing for 1200 h and by unstuffing for about 100 h. For a lens

18.8 mm long, aberration on the center axis is within 50 μm, the value of g is 0.050 mm^{-1} and the numerical aperture is 0.44. A rod lens of ~16 mm diameter has also been made by this technique [43, 44]. Figure 2.6 shows photographs using GRIN rod lenses of 9.4 mm and 16 mm diameter. The chromatic problem still remains as in the case of the ion-exchange process. The Abbe numbers were 25.4 at the center and 32.4 at the periphery of a rod made by double ion-exchange process.

2.4.2 Molecular stuffing technique

A molecular stuffing technique has been developed in principle for fabricating fibers [45] and has been applied to the fabrication of r-GRIN lenses [23, 24, 46]. The process consists of depositing an index-modifying dopant from the solution into a porous glass preform prepared by using the phase-separation and leaching process. The porous glass rod is first immersed in the aqueous solution of index-modifying dopant. The solution is allowed to diffuse into the pores until the concentration of the index-modifying dopant is uniform (the stuffing process). CsNO$_3$ or TlNO$_3$ is usually used as a refractive-index-modifying dopant because of its high refractive index and its high solubility in water. The rod preform with the dopant in its porous structure is then soaked in a solvent such as 40 vol% ethanol–water mixture at 70 °C (the unstuffing process) and the concentration profile of the dopant is changed along the radius by redissolution of the modifying dopant. The rod preform is then immersed

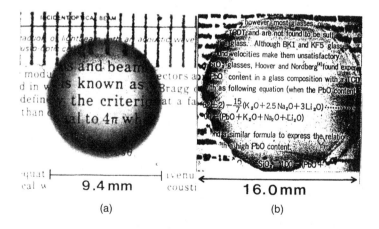

9.4 mm   16.0 mm

(a)   (b)

Fig. 2.6.   Photographs taken using GRIN rod lenses. (a) 9.4 mm diameter and (b) 16.0 mm diameter. [(a) reprinted from S. Ohmi, H. Sakai, Y. Asahara, S. Nakayama, Y. Yoneda and T. Izumitani, *Appl. Opt.* **27** (1988) 496, copyright (1988) with permission from the Optical Society of America.]

into a solution which has a low solubility for the modifying dopant such as ethanol or methanol at 0 °C so that the remaining dopant in the pores is precipitated and stabilized (the precipitating process). Finally, the doped pre-form is dried at a low temperature to remove any trapped solvent and heated at 850 °C to collapse the pore structure, which produces a solid glass rod (the consolidation process).

Fitting the index distribution, for instance in the case of TINO$_3$-doped rod of 3.4 mm in diameter, in Eq. (2.3) by the least squares technique, the optimum unstuffing time is found to be 7.5 min. Relating the optimum unstuffing time to the theoretically expected diffusion parameter, the diffusion coefficient of the dopant can be calculated. The results at 70 °C are: $D(TINO_3) = 8 \times 10^{-6}$ cm^2/s and $D(CsNO_3) = 5 \times 10^{-6}$ cm^2/s. These values are much larger than those of Tl$^+$ and Ag$^+$ at the glass transition temperature. GRIN lenses with a large diameter of ∼5 mm have been made using this technique. Although the effective region of this lens is limited to around 60% of its radius, the standard deviation is of the order of 10^{-5} in this region. Uni-directional gradient index slab lenses (2.5 ∼ 3.4 mm in thickness) with parabolic index profiles have also been fabricated by using this process and are expected to be used to produce a multiport LD module [24]. The disadvantages of this method are that the procedure is rather complicated and the unstuffing time is too short to control the concentration profile precisely.

2.4.3 Sol-gel technique

The fabrication of r-GRIN elements by the sol-gel process has been studied over recent years using two different methods to form the refractive index profile [47]: (1) partial leaching of index-modifying cations from an alkoxy-derived wet gel; and (2) interdiffusion of index-modifying cations in the liquid-filling micropores of a gel that is formed using an aqueous metal salt solution as the source of cations.

2.4.3.1 Partial leaching technique

A fabrication process of GRIN rod lenses by the partial leaching technique consists of the following step [48, 49]: (1) formation of a cylindrical wet gel from alkoxides of Si, Ti, Al, Zr, Ge, etc. in a plastic mold; (2) selective leaching of index-modifying cations such as Ti and Ge, by soaking the gel in a dilute acid; (3) washing and drying of the wet gel; and (4) sintering of the dry gel to glass. The typical fabrication time to obtain r-GRIN rods of a few millimetres in diameter is about 2 weeks. This includes leaching for 5 ∼ 15 h to form the compositional gradient of the index-modifying cations. The diffusion coefficient of the index-modifying ions is estimated to be 4×10^{-7} cm^2/s during the index gradient formation. These values are much larger than those in the ion-exchange process.

Rods of, for instance, about 2 mm in diameter, 20 mm in length and with a difference of refractive index between the center and the periphery Δn of about 0.02 have been fabricated using this process [50]. The rods have a smooth and nearly parabolic index gradient, which can be approximated as [48, 49, 51]

$$n(r) = n_0[1 - (r/r_0)^{1.94\sim2.35}]. \tag{2.20}$$

The resolution of the lens was calculated to be higher than 100 lines mm^{-1}. The GRIN lens shows a high coupling efficiency with optical fibers as well as good beam collimation properties [52].

GRIN rods of $3 \sim 7$ mm in diameter were also fabricated by the partial leaching technique from alkoxy-derived gels of ternary and quarternary silicate systems containing TiO_2 or ZrO_2 as the primary refractive index modifier [53]. The introduction of alkali into the TiO_2–SiO_2 system is effective in eliminating the crystallization and enables the incorporation of a large amount of Ti, which is essential to the fabrication of GRIN rods with a large value of Δn. The possibility of fabricating a GRIN rod with Δn as high as 0.08 has also been suggested [47].

The stress attributed to the less dense structure of the gel nearer the edge than at the center is inherent in the leaching method and becomes larger as the Δn increases. This stress can be relaxed, in principle, if the glass is annealed at a higher temperature for a longer time but the stress relaxation is very slow when the viscosity of the glass is very high and often remains uncompleted in an alkali-free high silica glass. Assuming the mechanical properties of the alkali titanium silicate system, the fabrication of GRIN rods with a diameters as large as 25 mm should be possible, if the cooling is carefully carried out after sufficient removal of residual stress [47].

2.4.3.2 Interdiffusion technique

Preparation of GRIN rods by the interdiffusion of index-modifying cations has been studied over the past 10 years [54–60]. The flow chart of a process is illustrated in Fig. 2.7 [47]. A wet gel of the PbO–B_2O_3–SiO_2 system is prepared from the precursor solution consisting of tetramethoxysilane (TMOS), tetraethoxysilane (TEOS), boronethoxide, and aqueous lead acetate solution. After being aged for a few days the wet gel is treated first with an aqueous solution of lead acetate, an isopropanol and water solution to remove acetic acid, and then soaked in the isopropanol–acetone system for precipitation of lead acetate in microcrystallites on the micropore wall of the wet gel. The gel

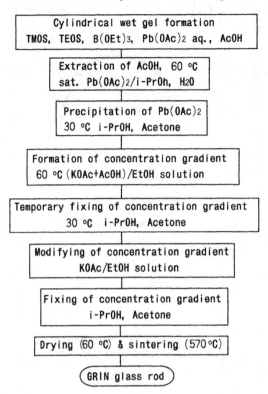

Fig. 2.7. Fabricating process of GRIN rods by sol-gel process based on the interdiffusion of Pb^{2+} and K^+ ions. [Reprinted from M. Yamane and S. Noda, *J. Ceram. Soc. Jpn.* **101** (1993) 11, copyright (1993) with permission from the Ceramic Society of Japan.]

thus treated is next soaked in an ethanolic solution of potassium acetate to form the compositional gradient within the gel. The compositional gradient is temporarily fixed by replacing the acetone by ethanol in the micropores and then modified by soaking again in the ethanolic solution of potassium acetate. After a final fixing of the compositional gradient, the gel is dried at 30 °C and sintered for densification at 570 °C.

The temporary fixing of the compositional gradient of index-modifying cations and another soaking of gel in the potassium acetate bath are necessary to modify the concentration gradient of lead near the periphery [55]. The process is further improved by replacing half of the potassium acetate by acetic acid during formation of the compositional gradient [59, 60]. The acetic acid controls the solubility of lead acetate within the gel during the subsequent treatment to fix the compositional profile with the isopropanol-acetone system solution and hinders the diffusion of lead ions in the vicinity of the periphery of the gel.

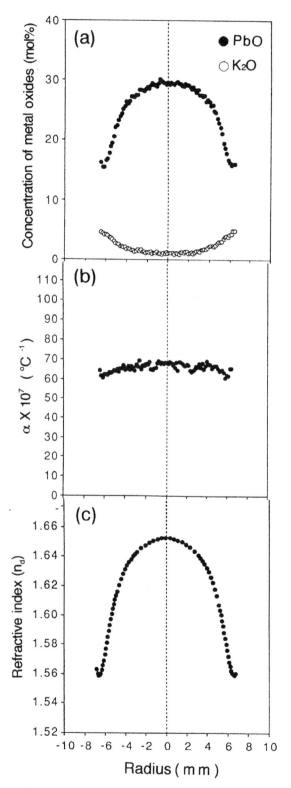

Fig. 2.8. Caption opposite.

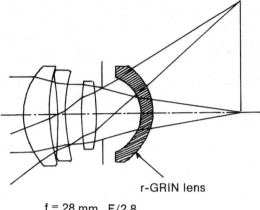

r-GRIN lens

f = 28 mm, F/2.8

Fig. 2.9. Composition of optical system for a practical test of the performance of a GRIN lens on a compact camera. [Reprinted from M. Yamane, H. Koike, Y. Kurasawa and S. Noda, *SPIE* vol. **2288**, Sol-Gel Optics III (1994) 546, copyright (1994) with permission from the Society of Photo-Optical Instrumentation Engineers.]

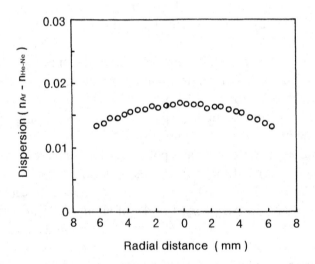

Fig. 2.10. Profile of refractive index dispersion in the GRIN rod of $PbO-K_2O-B_2O_3-SiO_2$ system. [Reprinted from M. Yamane and S. Noda, *J. Ceram. Soc. Jpn.* **101** (1993) 11, copyright (1993) with permission from the Ceramic Society of Japan.]

Fig. 2.8. Controlled concentration profiles of lead and potassium ions (a), thermal expansion coefficient (b) and refractive index profile (c) in the GRIN rod fabricated by the sol-gel interdiffusion technique. [Reprinted from M. Yamane, H. Koike, Y. Kurasawa and S. Noda, *SPIE* vol. **2288**, Sol-Gel Optics III (1994) 546, copyright (1994) with permission from the Society of Photo-Optical Instrumentation Engineers.]

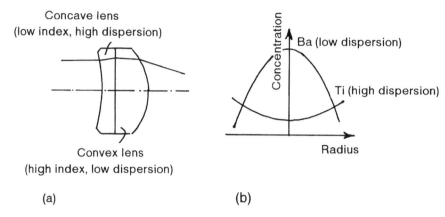

Fig. 2.11.   Design of a conventional achromatic lens (a) and a design to achieve a concentration distribution having a low dispersion distribution (b). [Reprinted from H. Kinoshita, M. Fukuoka, Y. Morita, H. Koike and S. Noda, *Proceedings SPIE* **3136**, Sol-Gel Optics IV (1997) 230, copyright (1997) with permission from the Society of Photo-Optical Instrumentation Engineers.]

GRIN rods of about 13 mm in diameter and 25 mm in length can be fabricated by this process starting with the formation of a wet gel of about 35 mm in diameter and 50 mm in length from the precursor sol of nominal composition $26PbO-7B_2O_3-67SiO_2$ (mol%). From the optimum soaking time of $8 \sim 16$ h for the formation of the compositional gradient, the diffusion coefficient is calculated to be $0.7 \sim 1.5 \times 10^{-6}$ cm^2 s^{-1}, which is very much larger than those of the index-modifying ions in the ion-exchange process but smaller than in the molecular stuffing process.

The concentration profile of lead and potassium ions and the refractive index profile along the diameter of the rods are respectively shown in Fig. 2.8 [60]. The smooth and large change in refractive index over nearly the entire diameter from the center toward the periphery, i.e. Δn of about 0.095 and the effective diameter as large as 12 mm, is favourable for application as an imaging lens element.

For the fabrication of GRIN rods with a larger Δn and larger diameter, it is also necessary to control the thermal expansion coefficient in the radial direction, as mentioned in Section 2.3.4. According to A. A. Appen [61], PbO is the major factor which influences the refractive index, whereas K_2O has the largest effect on the thermal expansion coefficient. As shown in Fig. 2.8(a), the gradient of the potassium ion is much smaller than expected from the lead concentration profile. Owing to this decrease in potassium concentration near the periphery, the thermal expansion coefficient of the material decreases and becomes almost constant throughout the diameter of the rod as shown in

Table 2.2. *Comparison of various techniques for fabricating r-GRIN glass materials*

Method	Merit	Demerit	Matrix materials	Dopant ions	Maximum diameter, r_0 (mm)	Time for $n(r)$ formation, t_0 (h)	Optical properties	Applications
Ion exchange $(T < T_g)^a$	Precise control of gradient index	Long manufacturing time	Alkali-boro-silicate glasses	Tl^+, Cs^+	4.0	250	Δn:0.11 NA:0.62	Lens array for copy machines and facsimiles
b		Small diameter		Tl^+: Ag^+ Ag^+: Tl^+	$(D = 10^{-9} \sim 10^{-10}\ \mathrm{cm^2/s})$ 4.0 $(D = 10^{-8}\ \mathrm{cm^2/s})$	72		Pick-up lenses for CD Connector and coupler for optical fiber communications
$(T > T_g)^c$					20	412		
Ion stuffing^d	Comparatively fast $n(r)$ formation	Little complicated procedure	Phosphate	Ag^+	16 $(D = 10^{-8}\ \mathrm{cm^2/s})$	1200	Δn:0.06 NA:0.6	Objective lens for camera
Molecular stuffing^e	Thermo and chemical resistive Fast $n(r)$ formation	Complicated procedure	Silica-rich	Tl^+, Cs^+	5.0 $(D = 10^{-6}\ \mathrm{cm^2/s})$	< 1	Δn:0.025 NA:0.3	Slab lens devices for optical communications
Sol-gel process Leaching^f	Fast $n(r)$ formation	Shrinkage during procedure	Silica and silica-rich glasses	Ge^{4+}, Ti^{4+}	7 $(D = 10^{-7}\ \mathrm{cm^2/s})$	< 20	Δn:0.02 NA:0.2	Coupler for optical communications
Inter-diffusion^g	Fast $n(r)$ formation Dispertion control (Ion selectable)	Shrinkage during procedure	Lead boro-silicate glasses	Pb^{2+}	13 $(D = 10^{-6} \sim 10^{-7}\ \mathrm{cm^2/s})$	$8 \sim 16$	Δn:0.095	Lens for zoom lens, compact camera and video cameras
CVD process^h	Simple process	unsmooth $n(r)$	Silica	Ge^{4+}	2		Δn:2%	Coupler for optical communications

NA, numerical aperture [a] [8, 17, 31–36]; [b] [38–40]; [c] [37]; [d] [41–44]; [e] [23, 24, 45, 46]; [f] [47–53]; [g] [27–30, 47, 54–60]; [h] [62].

Fig. 2.8(b). The fabrication of GRIN rods of a PbO–K$_2$O-based system with diameter of 20 ∼ 25 mm should be possible by optimizing the sintering and cooling conditions. The performance test of the lens on a compact camera that is shown in Fig. 2.9 demonstrated that the refractive index profile of the GRIN rod is satisfactory for application in practical optical systems.

Figure 2.10 shows the profile of refractive index dispersion, $n_{Ar} - n_{He-Ne}$, determined for the wavelength of an Ar laser (488 nm) and a He–Ne laser (633 nm) in a GRIN rod of the PbO–K$_2$O–B$_2$O$_3$–SiO$_2$ system [47]. The large decrease in dispersion from the center towards the periphery is due to the changing lead concentration. As already mentioned in Section 2.3.3, the chromatic aberration of the r-GRIN lenses can be minimized if a parabolic refractive index profile can be formed by a convex distribution of low dispersion ions (e.g. Ba^{2+}) and a concave distribution of high dispersion ions (e.g. Ti^{4+}), in the same way as for the combination of an achromatic lens (Fig. 2.11(a) and (b)) [27–30]. A GRIN fabrication that is based on the ion pairs of such multivalent ions can only be carried out using the sol-gel interdiffusion technique, in which the formation of a concentration profile of index-modifying cations is made in the precursor gel regardless of the valence of the cations. Most recently, r-GRIN materials with a low dispersion distribution were successfully fabricated using the ion pair of divalent Ba^{2+}–Ti^{4+} in a sol-gel process [27, 29]. This lens has a dispersion distribution in the low dispersion region ($V < V_{10}$), which should be particularly useful for the lens (effective diameter: 4 mm) in video cameras [28, 30].

2.4.4 Comparison between various fabrication techniques

Table 2.2 summarizes the merits, demerits and some other features of the various fabrication techniques that are used to make r-GRIN glass materials.

2.5 Spherical gradient index lens

One example of a practical spherical gradient lens is a planar micro-lens used in two-dimensionally arrayed devices, which is fabricated using the following steps. A thin metal mask preventing the diffusion of index-modifying ions is first coated on the substrate and then many small apertures of submillimeter diameter are patterned on it using a photolithography technique. Next the substrate is immersed in a molten salt bath at high temperatures. In this process, index-modifying ions in the molten salt, for example, Ag$^+$ or Tl$^+$, diffuse into the glass substrate through the mask aperture and form a hemispherical distribution for the refractive index [63].

If the aperture is small enough compared with the diffusion length, the aperture acts as a small diffusion source with radius r_m. The dopant concentration at time t is given by the equation [20]

$$C(r, t) = C_0(r_m/r)\text{erfc}\{(r - r_m)/2(Dt)^{1/2}\} \qquad (2.21)$$

where r is the distance from the center of the diffusion source, and D is the diffusion constant. The preliminary experiment was made on a planar BK7 glass and Tl^+ molten salts [63]. The depth of the gradient index region was however, not large ($\sim$0.4 mm) compared with the lens diameter.

The electro-migration technique is effective in making a lens that has a large diffusion depth in a reduced fabrication time. In this technique, the electric field $E = r_m V_s$ is applied to the substrate in the thickness direction and the ionic flow created by the electric field plays an important role in the formation of the refractive index distribution. The ion concentration is then given by [64]

$$C(r, t) = C_0 h[t - (r^3 - r_m^3)/(3\mu r_m V_s)] \qquad (2.22)$$

where, $V_s = (2/\pi)V_0$, V_0 is the electric potential on the mask aperture, μ is the ion mobility, C_0 is the concentration of ions on the source surface and h is a step function: $h(x) = 1$ for $x > 0$ and 0 for $x < 0$. The gradient index is expressed by the sum of the functions of r (radial direction) and z (axial direction), and is given for a parabolic index profile by [65]

$$n^2(r, z) = n^2(0, 0)\{1 - (gr)^2 - v_{20}(gz)^2\}(0 < z < d) \qquad (2.23)$$

where v_{20} is the constant related to the three-dimensional index distribution [16].

A current attractive application of this type of planar micro-lens array is color LCD projectors, in which the brightness of the projected image is enhanced by a focusing effect from each lenslet [66]. This new system improves the brightness more than three times in comparison with a conventional system.

References

1 E. W. Marchand: *Gradient Index Optics*, (Academic Press Inc., New York, 1978).
2 D. T. Moore: *Appl. Opt.* **19** (1980) 1035.
3 D. T. Moore: *Ceramic Bulletin* **68** (1989) 1941.
4 I. Kitano, H. Ueno and M. Toyama: *Appl. Opt.* **25** (1986) 3336.
5 K. Nishizawa and H. Nishi: *Appl. Opt.* **23** (1984) 1711.
6 H. Nishi, H. Ichikawa, M. Toyama and I. Kitano: *Appl. Opt.* **25** (1986) 3340.
7 K. Kikchi, S. Ishihara, H. Shimizu and J. Shimada: *Appl. Opt.* **19** (1980) 1076.
8 T. Miyazawa, K. Okada, T. Kubo, K. Nishizawa, I. Kitano and K. Iga: *Appl. Opt.* **19** (1980) 1113.
9 M. Daimon, M. Shinoda and T. Kubo: *Appl. Opt.* **23** (1984) 1790.
10 W. J. Tomlinson: *Appl. Opt.* **19** (1980) 1127.

11 M. Kawazu and Y. Ogura: *Appl. Opt.* **19** (1980) 1105.
12 K. Uehira and K. Koyama: *Appl. Opt.* **29** (1990) 4081.
13 L. G. Atkinson, S. N. Houde-Walter, D. T. Moore, D. P. Ryan and J. M. Stagaman: *Appl. Opt.* **21** (1982) 993.
14 N. Aoki, T. Nagaoka and H. Tsuchida: *Proc. 16th Symp. Opt. Soc. Jpn.*, June (The Optical Society of Japan, Tokyo, 1991) p. 39.
15 K. Iga: *Appl. Opt.* **19** (1980) 1039.
16 K. Iga, Y. Kokubun and M. Oikawa: *Fundamentals of Microoptics, Distributed-Index, Microlens and stacked Planar Optics*, Chapter 2. (Academic Press, Inc. Tokyo, 1984) (OHM, Tokyo–Osaka–Kyoto, 1984) p. 8.
17 H. Kita, I. Kitano, T. Uchida and M. Furukawa: *J. Amer. Ceram. Soc.* **54** (1971) 321.
18 E. W. Marchand: *Appl. Opt.* **19** (1980) 1044.
19 I. Fanderlik: Optical Properties of Glass, *Glass Science and Technology 5*, Chapter 4, (Elsevier, Amsterdam, 1983) p. 78.
20 J. Crank: *The Mathematics of Diffusion*, 2nd Edn. Chapters 4, 5, and 6, (Clarendon Press, Oxford, 1975) p. 69.
21 I. Kitano, K. Koizumi, H. Matsumura, T. Uchida and M. Furukawa: *Jpn. J. Appl. Phys.* (supplement) **39** (1970) 63.
22 W. E. Martin: *Appl. Opt.* **14** (1975) 2427.
23 Y. Asahara and S. Ohmi: *Technical Digest, Fourth Topical Meeting on Gradient-Index Optical Imaging Systems*, Monterey (1984) ThE-B1.
24 Y. Asahara, H. Sakai, S. Shingaki, S. Ohmi, S. Nakayama, K. Nakagawa and T. Izumitani: *Appl. Opt.* **24** (1985) 4312.
25 J. R. Tessman, A. H. Kahn and W. Shockley: *Phys. Rev.* **92** (1953) 890.
26 S. D. Fantone: *Appl. Opt.* **22** (1983) 432.
27 H. Kinoshita, M. Fukuoka, Y. Morita, H. Koike and S. Noda: *Proc. SPIE* **3136** *Sol-Gel Optica IV*, (Society of Photo-Optical Instrumentation Engineers, Bellingham, 1997) 230.
28 H. Tsuchida, T. Nagaoka and K. Yamamoto: *Technical Digest, The Sixth Microoptics Conference and the Fourteenth Topical Meeting on Gradient-index Optical Systems*, (The Japan Society of Applied Physics, Tokyo, 1997) p. 32.
29 Y. Morita, M. Fukuoka, H. Koike, H. Kinoshita and S. Noda: *ibid*, p. 398.
30 K. Yamamoto, H. Tsuchida and S. Noda: *Proceedings, International Workshop on Optical Design & Fabrication*, Tokyo (1998) p. 25.
31 T. Yamagishi, K. Fujii and I. Kitano: *Appl. Opt.* **22** (1983) 400.
32 T. Yamagishi, K. Fujii and I. Kitano: *J. Non-Cryst. Solids* **47** (1982) 283.
33 K. Fujii and N. Akazawa: *Summaries of Topical Meeting on Gradient-Index Optical Imaging Systems*, (The Optical Society of America, Reno, 1987) p. 102.
34 D. S. Kindred, J. Bentley and D. T. Moore: *Appl. Opt.* **29** (1990) 4036.
35 K. Fujii, S. Ogi and N. Akazawa: *Appl. Opt.* **33** (1994) 8087.
36 Optical guide of Nippon Sheet Glass *Selfoc Micro Lens*.
37 Y. Kaite, M. Kaneko, S. Kittaka, M. Toyama and I. Kitano: *Proc. Annual Meeting Jpn. Soc. Appl. Phys.* (in Japanese), (1989) 2a-Z-4.
38 D. S. Kindred: *Appl. Opt.* **29** (1990) 4051.
39 J. B. Caldwell, D. S. Kindred and D. T. Moore: *Summaries of Topical Meeting on Gradient Index Optical Imaging Systems*, (The Optical Society of America, Reno, 1987) p. 80.
40 J. L. Coutaz and P. C. Jaussaud: *Appl. Opt.* **21** (1982) 1063.

41 S. Ohmi, H. Sakai and Y. Asahara: *Summaries of Topical Meeting on Gradient-Index Optical Imaging Systems*, (The Optical Society of America, Reno, 1987) p. 83.
42 S. Ohmi, H. Sakai, Y. Asahara, S. Nakayama, Y. Yoneda and T. Izumitani: *Appl. Opt.* **27** (1988) 496.
43 S. Ohmi, H. Sakai and Y. Asahara: *Technical Digest of 1st Micro-optics Conference*, Tokyo, (Japan Society of Applied Physics, Tokyo, 1987) p. 40.
44 Y. Asahara and A. J. Ikushima: *Optoelectronics – Devices and Technologies*, **3** (1988) 1.
45 J. H. Simmons, R. K. Mohr, D. C. Tran, P. B. Macedo and T. A. Litovitz: *Appl. Opt.* **18** (1979) 2732.
46 R. K. Mohr, J. A. Wilder, P. B. Masedo and P. K. Gupta: *Technical Digest, Gradient Index Optical Imaging Systems*, Rochester (1979) WA1-1.
47 M. Yamane and S. Noda: *J. Ceram. Soc. Jpn.* **101** (1993) 11, and refs cited therein.
48 K. Shingyouchi, S. Konishi, K. Susa and I. Matsuyama: *Electronics Lett.* **22** (1986) 99.
49 K. Shingyouchi, S. Konishi, K. Susa and I. Matsuyama: *Electronics Lett.* **22** (1986) 1108.
50 S. Konishi: *SPIE* **1328** 'Sol-Gel Optics' (1990) 160.
51 S. Konishi, K. Shingyouchi and A. Makishima: *J. Non-Cryst. Solids* **100** (1988) 511.
52 K. Shingyouchi and S. Konishi: *Appl. Opt.* **29** (1990) 4061.
53 T. M. Che, J. B. Caldwell and R. M. Mininni: *SPIE*, **1328** 'Sol-Gel Optics' (1990) 145.
54 M. Yamane, J. B. Caldwell and D. T. Moore: *J. Non-Cryst. Solids* **85** (1986) 244.
55 M. Yamane, H. Kawazoe, A. Yasumori and T. Takahashi: *J. Non-Cryst. Solids* **100** (1988) 506.
56 M. Yamane, A. Yasumori, M. Iwasaki and K. Hayashi: *Proc. Mater. Res. Symp.*, **180** (1990) 717.
57 M. Yamane, A. Yasumori, M. Iwasaki and K. Hayashi: *SPIE* **1328** 'Sol-Gel Optics' (1990) 133.
58 M. Yamane and M. Inami: *J. Non-Cryst. Solids* **147 & 148** (1992) 606.
59 M. Maeda, M. Iwasaki, A. Yasumori and M. Yamane: *J. Non-Cryst. Solids* **121** (1990) 61.
60 M. Yamane, H. Koike, Y. Kurasawa and S. Noda: *SPIE* **2288** 'Sol-Gel Optics III' (1994) 546.
61 A. A. Appen: *Chemistry of Glass* (Nissotsushinsha, Tokyo, 1974) p. 318.
62 K. Nakatate, N. Shamoto, T. Oohashi and K. Sanada: *SPIE* **2131** (1994) 203.
63 M. Oikawa and K. Iga: *Appl. Opt.* **21** (1982) 1052.
64 M. Oikawa, K. Iga, M. Morinaga, T. Usui and T. Chiba: *Appl. Opt.* **23** (1984) 1787.
65 K. Iga, M. Oikawa and J. Banno: *Appl. Opt.* **21** (1982) 3451.
66 K. Koizumi: *Proc. International Workshop on 'Advanced Materials for Multifunctional Waveguides'* Chiba (1995) p. 8.

3

Laser glass

Introduction

The laser is a source of monochromatic radiation of high intensity, coherence, and directionality, in the ultraviolet, visible and infrared optical region. The many and varied applications of lasers include laboratory use, research (optical spectroscopy, holography, laser fusion), materials-processing (cutting, scribing, drilling, welding, abrasion), communications, information processing, military (range finders, target designators), and medical use [1]. Glass plays many varied roles in rare-earth laser systems, because glass can be made with uniformly distributed rare-earth concentrations and has great potential as a laser host medium. In addition, rare-earth doped fibers have received growing attention recently. They could have many uses as amplifiers in optical communication systems and as optical sources. Glass waveguide lasers are another interesting subject for the development of compact laser sources and amplifier devices.

This chapter concentrates on laser glass materials containing rare-earth ions, thus excluding crystalline laser materials. To provide an understanding of the properties of laser glass, the chapter begins with a brief summary of the fundamental physics of lasers in Section 3.1. To explain the characteristics of bulk laser glass, representative laser parameters and their host dependences are summarized briefly in Section 3.2. The recent developments of fiber lasers (in Section 3.3) and glass waveguide lasers (in Section 3.4) are reviewed.

3.1 Fundamentals of laser physics

3.1.1 Stimulated emission [2, 3]

It is well known that atoms, ions and molecules can exist in certain stationary states, each of which is characterized by levels of the atomic system called quantum numbers. To simplify the discussion, let us consider an idealized system as shown in Fig. 3.1 with just two nondegenerate energy levels, 1 and 2, having populations N_1 and N_2, and energies E_1 and E_2, respectively. The total number of atoms in these two levels is assumed to be constant

$$N_1 + N_2 = N. \tag{3.1}$$

If a quasimonochromatic electromagnetic wave of frequency ν_{21} passes through an atom system with energy gap $h\nu_{21}$,

$$h\nu_{21} = E_2 - E_1 \tag{3.2}$$

the atoms can transfer from level 1 to level 2 by absorbing energy and then the population at the lower level will be depleted at a rate proportional both to the radiation density $\rho(\nu)$ and to the population N_1 at that level,

$$(dN_1/dt) = -B_{12}\rho(\nu)N_1 \tag{3.3}$$

where B_{12} is a constant of proportionality. The product $B_{12}\rho(\nu)$ thus can be interpreted as the transition rate from level 1 to level 2 per atom and per unit frequency.

After an atom has been raised to the upper level 1 by absorption, it will return to the lower level 2 and emit energy via two distinct processes [4]: (1) radiative decay with the emission of a photon; (2) nonradiative decay where the excitation energy is converted into vibrational quanta of the surroundings. The observed lifetime or fluorescence lifetime τ_f at an excited level 2 is given by the combination of probabilities for radiative (A_{rad}) and nonradiative (W_{nr}) processes.

$$\begin{aligned}(1/\tau_f) &= A_{rad} + W_{nr} \\ &= (1/\tau_R) + W_{nr}\end{aligned} \tag{3.4}$$

where τ_R is the radiative lifetime. The nonradiative probability W_{nr} includes relaxation by multiphonon emission and effective energy transfer rates arising from ion–ion interactions. The radiative quantum efficiency η is defined by

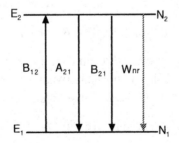

Fig. 3.1.   Simple two-level system with population N_1, N_2 and energy E_1 and E_2 respectively.

$$\eta = A_{rad}/(A_{rad} + W_{nr})$$
$$= \tau_f/\tau_R. \tag{3.5}$$

The relative probabilities for radiative or nonradiative decay between given levels may range from comparable values to the two extremes where $A_{rad} \ll W_{nr}$ or $A_{rad} \gg W_{nr}$.

If the transitions from level 2 to level 1 take place only via a radiative process, the population of the upper level 2 decays spontaneously to the lower level at a rate proportional to the upper level population (spontaneous emission) or it decays under stimulation by external electromagnetic radiation of an appropriate frequency (stimulated emission). The transition rate for spontaneous emission and stimulated emission from level 2 to level 1 is respectively given by

$$(dN_2/dt) = -A_{21}N_2 \tag{3.6}$$
$$(dN_2/dt) = -B_{21}\rho(v)N_2 \tag{3.7}$$

where A_{21} and B_{21} are the transition probabilities and are known as Einstein's A and B coefficient respectively, and $\rho(v)$ is the external radiation density per unit frequency. For spontaneous emission, there is no phase relationship between the individual emission processes. Equation (3.6) has a solution

$$N_2(t) = N_2(0)\exp(-t/\tau_{21}) \tag{3.8}$$

where τ_{21} is the radiative lifetime of level 2 which is equal to the reciprocal of the A coefficient

$$\tau_{21} = 1/A_{21}. \tag{3.9}$$

In stimulated emission, the phase is the same as that of the stimulating external radiation, that is, the external radiation gains the stimulated radiation and the material acts as an amplifier of radiation at the proper frequency $v_{21} = (E_2 - E_1)/h$. This is the basic phenomenon of laser action. In the presence of external radiation, we must combine the spontaneous and stimulated emission, and the total transition probability downward is

$$(dN_2/dt) = -\{A_{21} + B_{21}\rho(v)\}N_2. \tag{3.10}$$

Comparing this expression for the radiation density in thermal equilibrium with the black body radiation law, the relations between the A and B coefficients are usually given in a form that is known as Einstein's relations.

$$(A_{21}/B_{21}) = (8\pi h v^3/c^3) \tag{3.11}$$
$$B_{21} = B_{12}. \tag{3.12}$$

3.1.2 Cross section [3]

In the previous section, it is assumed that a monochromatic wave with frequency ν_{21} acts on a two-level system with an infinitely sharp energy gap $h\nu_{21}$. An actual atomic system, however, has a finite transition linewidth $\Delta\nu$ with center frequency ν_0 and a signal has a bandwidth $d\nu$ with center frequency ν_s. The latter is usually narrow compared with the former. Laser materials, therefore, are generally characterized as the interaction of linewidth-broadened energy levels with a monochromatic wave. Then, introducing a parameter 'cross section', $\sigma_{21}(\nu_s)$, Einstein's coefficient for the radiative transition from level 2 to level 1 is replaced by the form

$$\sigma_{21}(\nu_s) = (h\nu_s g(\nu_s, \nu_0)/c)B_{21}$$
$$= (A_{21}\lambda_0^2/8\pi n^2)g(\nu_s, \nu_0) \qquad (3.13)$$

where $g(\nu, \nu_0)$ is the atomic lineshape function. The cross section is a very useful parameter which we will use to characterize the laser materials in the following section.

The lineshape of an atomic transition depends in the cause of line broadening, leading to two distinctly different mechanisms, homogeneous and inhomogeneous broadening. As glass is a disordered medium there is a site-to-site variation in the splitting of the rare-earth energy levels, which produces inhomogenous line broadening of the same order of magnitude as the spacing between the individual Stark levels ($\sim 100~\text{cm}^{-1}$). The inhomogeneous-broadening linewidth can be represented by a Gaussian distribution. The peak value of the normalized Gaussian curve is

$$g(\nu_0) = (2/\Delta\nu)(\ln 2/\pi)^{1/2}. \qquad (3.14)$$

If we assume $\nu \sim \nu_s \sim \nu_0$, the spectral stimulated emission cross section at the center of the atomic transition is given by

$$\sigma_{21}(\nu_s) = (A_{21}\lambda_0^2/4\pi n^2\Delta\nu)(\ln 2/\pi)^{1/2}. \qquad (3.15)$$

3.1.3 Creation of a population inversion [2, 3]

3.1.3.1 Population inversion

From Eq. (3.12), the possibility for stimulated absorption or stimulated emission depends on which population, i.e. N_1 or N_2, is larger than the other. According to the Boltzman distribution, a level 1 with lower energy is more densely populated than a level 2 with a higher energy at any temperature. Therefore the population difference $\Delta N = N_2 - N_1$ is always negative, which means that the absorption coefficient α is positive and the incident radiation is absorbed. Thus a beam passing through a material will always lose intensity.

If it is possible to achieve a temporary situation such that there are more atoms in an upper energy level than in a lower energy level, i.e. $N_2 > N_1$, the normal positive population difference on that transition then becomes negative (population inversion), and the normal absorption of an applied signal on that transition is correspondingly changed to stimulated emission, or amplification of the applied signal. The maximum gain generated by the population inversion ΔN can be expressed by

$$g = \Delta N \sigma_{21}(\nu_s). \tag{3.16}$$

The main features are now how the necessary population inversion is obtained in a solid. The laser operations of an actual laser material typically involve a very large number of energy levels with complex excitation processes and relaxation processes among all these levels. These can be understood, however, by using the simplified three-level or four-level energy diagrams as shown in Fig. 3.2.

3.1.3.2 Three-level system

Initially, all atoms of the laser materials are at the ground level 1. The pump light raises these atoms from the ground level into the broad band 3 (known as optical pumping). Most of the excited atoms decay rapidly and nonradiatively into the intermediate level 2. Finally, the atoms return to the ground level by the emission of a photon. This last transition is responsible for the laser action.

As the terminal level of the laser transition is the highly populated ground state in this system, more than half of the atoms in the ground level must be excited to the metastable level 2 in order to maintain a specified amplification, that is,

$$N_2 > N_1 = N/2. \tag{3.17}$$

There is the additional requirement that the life-time of level 2 should also be large in comparison with the relaxation time of the $3 \sim 2$ transition, i.e. $\tau_{21} \gg \tau_{32}$.

3.1.3.3 Four-level system

To overcome the disadvantages of the three-level system, it is necessary to utilize a four-level system. There is in this case an additional normally unoccupied level above ground level at which the relevant transitions terminate.

As in the case of the three-level system, the atoms that are excited from the ground level 0 to level 3 will decay rapidly and nonradiatively to level 2. The

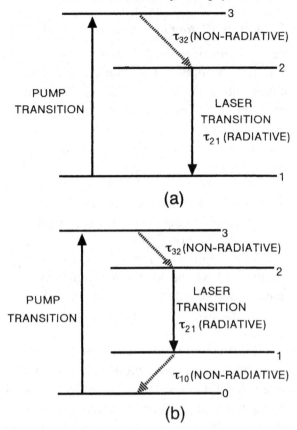

Fig. 3.2. Simplified energy diagrams used to explain the operation of optically pumped (a) three-level and (b) four-level laser systems.

laser transition takes place between the excited level 2 and the terminal level 1. From the level 1 the atom decays by a rapid nonradiative transition to the ground level 0. The terminal level 1 will be empty because this level is far above the ground state so that its thermal population is small. In this system, the condition for maintaining amplification is simply

$$\Delta N = N_2 - N_1$$
$$= N_2 > 0. \tag{3.18}$$

In addition, in the most favorable case, the relaxation times of the $3 \sim 2$ and $1 \sim 0$ transitions are short compared with the spontaneous emission lifetime of the laser transition τ_{21}. In a four-level system, therefore, all pumping power is used to produce gain through ΔN and an inversion of the $2 \sim 1$ transition can occur even with small pump power. The thresholds for four-level lasers are typically lower than those of the three-level systems.

3.1.4 Laser oscillation by resonant cavity [1–3]

A typical laser oscillator is shown in Fig. 3.3. The optical cavity is constructed by two mirrors that reflect the optical radiation back and forth in the laser material, thereby implementing the optical feedback element. The pumping light inverts the population in the laser material, thereby giving the system a gain factor exceeding unity. The oscillation is first triggered by some spontaneous radiation emitted along the axis of the laser material. This radiation is amplified and when the mirror at the output end of the cavity is partially transparent, a fraction of the radiation will leak out or be emitted from the cavity and can be utilized externally as a laser beam. At the same time some radiation is reflected and amplified again as it passes through the laser material and continues to transverse the laser oscillator.

Assuming a gain per unit length of g in the laser material of length L and mirrors having a reflectivity of R_1 and R_2, the light intensity gains by a factor of $\exp(gL)$ in each passage through the material. In an oscillator, on the other hand, a number of loss mechanisms attenuate the light: reflection, scattering and absorption losses in the elements in the cavity such as the mirrors and amplifying medium, and diffraction losses. It is convenient to lump all of the nonoutput losses into a single parameter, the absorption coefficient per unit length α. After passing through the laser material, reflecting off the mirror R_1 and returning through the material to R_2 and reflecting, the loop gain on every complete two-way passage of the light through the laser cavity is given by

$$G = R_1 R_2 \exp[2(g - \alpha)L]. \tag{3.19}$$

Whether the light will be attenuated or strongly amplified depends on whether $g - \alpha < 0$ or $g - \alpha > 0$. The condition for continuing oscillation is then $G = 1$.

Assuming now that the pump rate is a linear function of input pump power

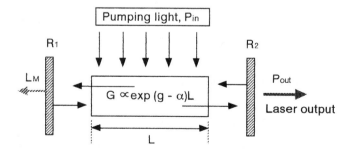

Fig. 3.3.   Typical laser oscillator configuration.

P_{in}, the population inversion is given by a function of the pump intensity. Output power versus input pump power for a four-level system is given schematically in Fig. 3.4. The following expression for the laser output power is then obtained

$$P_{out} = \sigma_s(P_{in} - P_{th}) \tag{3.20}$$

where P_{th} is the input pump power required to achieve threshold and σ_s the slope efficiency of the laser output versus input pump curve. The excess pump power beyond that required to provide the population inversion which equilibrates the losses contributes to the available output power P_{out}. The thresholds for three-level lasers are typically higher than those of the four-level systems.

3.1.5 Active ions for laser glasses

Materials for laser operation must possess sharp fluorescent lines, strong absorption bands, and a reasonably high quantum efficiency for the fluorescent transition of interest. These characteristics are generally shown by solids which incorporate a small amount of active ions in which the optical transitions can occur between states of inner incomplete electron shells. Trivalent lanthanide ions have been the most extensively used active ions, because there are many fluorescing states and wavelengths to choose from among the 4f electron configurations.

According to the quantum theory [5], the state of an electron in an atom is

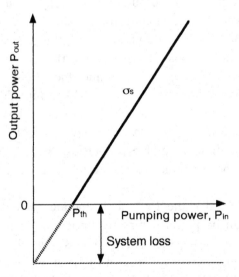

Fig. 3.4.   Output power versus pump input for a four-level system.

characterized by three orbital quantum numbers n, l, m and by a spin quantum number s. A single electron is described by a symbol using small letters according to the sequence: $l = 0, 1, 2, 3, \ldots$ is s, p, d, f, $\ldots$ respectively.

Rare-earth ions of the lanthanide series are characterized by a closed inner shell of the palladium configuration, with the additional configuration $4f^n5s^2p^6$. The screening effect of the 5s and 5p electrons makes the energy levels of the 4f electrons less sensitive to the surrounding ions. In such a multi-electron ion, the energy levels for electronic states arise from the combination of the Coulomb interaction among the electrons and the spin-orbit interaction, which provide complex and varied optical energy-level structures of ions. The splitting of these energy levels is usually labeled by the well known Russel-Sounders multiplet notation [6, 7], $^{2S+1}(L)_J$. Here J is the quantum number of the total angular momentum which is obtained by summing vectorially the total orbital angular momentum L and total spin angular momentum S. Thus, there are many possible three- and four-level lasing schemes. The large number of excited states that are suitable for optical pumping and subsequent decay to metastable states having high quantum efficiencies and narrow emission lines are also favorable for achieving laser action [8, 9]. All glass lasers to date have used trivalent lanthanides as the active ions. There are many books and articles in which the behavior of active ions in crystals and in glass have been summarized [1, 6, 8, 10–17].

Of the many rare-earth ions, Nd^{3+} is well known for its favorable character-istics for laser transition [1, 8, 10, 11]. As can be seen in Fig. 3.5.(a), the levels denoted $^2H_{9/12}$ or $^4F_{5/2}$ can be used to absorb light from a pumping source at about 800 nm. Ions excited into the pump band decay, usually nonradiatively, to the upper laser level $^4F_{3/2}$ in a time that is short in comparison with its radiative lifetime (typically $300 \sim 600$ µs). From this upper laser level $^4F_{3/2}$ there are four laser transitions $^4I_{15/2}$ (1.80 µm), $^4I_{13/2}$ (1.35 µm), $^4I_{11/2}$ (1.06 µm), and $^4I_{9/2}$ (0.88 µm). In particular the $^4F_{3/2} - ^4I_{11/2}$ (1.06 µm) transition has a high branching ratio ($0.45 \sim 0.48$) and in addition there is still a further nonradiative transition from the lower lasing level ($^4I_{11/12}$) to the ground state. This situation allows the description of Nd^{3+} as a four-level system. As a result of these favorable properties, Nd^{3+} is widely used as a dopant in a large number of bulk laser materials and for extensive practical applications.

As shown in Fig. 3.5.(b), the Er^{3+} ions operate as a three-level system [1, 8, 10, 11]. They can be pumped directly into the metastable upper lasing level $^4I_{13/2}$(1.49 µm) or into higher energy levels, e.g. $^4I_{11/2}$ (980 nm), $^4I_{9/2}$ (807 nm), $^4F_{9/2}$ (660 nm) or $^2H_{11/2}$ (514.5 nm), from which nonradiative decay occurs to the metastable $^4I_{13/2}$ level. Lasing as a three-level scheme can be

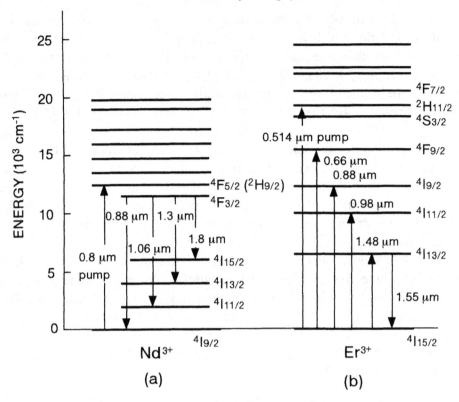

Fig. 3.5. Partial energy levels and lasing transitions for (a) Nd^{3+} and (b) Er^{3+}. (Energy levels from [18] and lasing transitions from [1, 8, 10, 11].)

obtained on the $^4I_{13/2} - {}^4I_{15/2}$ transition. The lower lasing level, in this case is either the ground state ($^4I_{15/2}$) or a level close to the ground state.

Absorption from the ground state directly to the upper lasing level takes place, causing competition with the emission of lasing photons. Er^{3+} lasers therefore have a higher power threshold than Nd^{3+} lasers.

3.1.6 Laser parameters and their host dependence [4, 13–16, 19, 20]

As mentioned above, the $4f^n$ electron state configuration of the rare-earth ions is not sensitive to the surrounding ions but the transition probability between these electron states is sensitive to the surrounding ions, both radiatively and nonradiatively.

The medium which concerns us in optics can be thought of as a collection of charged particles: electrons and ion cores. The elastic displacement of the electronic charge distribution to the nucleus results in a collection of induced electric-dipole moments. A light wave, on the other hand, consists of electric and magnetic fields which vary sinusoidally at optical frequency. The effect of the optical field, therefore, can be expressed classically in terms of the field-induced electric-dipole moment [21, 22]. Optical transitions in rare-earth ions also have been found to be predominantly electric-dipole in nature. For transitions between $4f^n$ electrons, no change of parity is involved and electric-dipole transitions are principally forbidden by Laporte's rule. In a crystal or glass matrix with crystal-fields which lack inversion symmetry at the active ion, however, the slight, crystal-field-induced admixture of other wave functions (such as from the 5d states) with the 4f wave functions creates a small probability for electric-dipole transitions. Magnetic dipole and electric quadrupole transitions are allowed by the selection rule but their contributions to radiative decay are generally small or negligible [13–16].

It has been shown independently by B. R. Judd and G. S. Ofelt that it is possible to make the calculation of electric-dipole transition probabilities tractable by assuming certain approximations [23, 24]. In this theory, the line strength of an $f - f$ transition in a rare-earth ion is expressed in terms of a sum of the products of three intensity parameters Ω_λ and squared matrix elements $U(\lambda)$ between the initial J-states $|(S, L)J\rangle$ and terminal J' state $|(S', L')J'\rangle$, of the form [13, 19, 20, 25]

$$S = \sum_{\lambda=2,4,6} \Omega_\lambda (U(\lambda))$$

$$= [\Omega_2 U(2) + \Omega_4 U(4) + \Omega_6 U(6)] \qquad (3.21)$$

$$U(\lambda) = |\langle (S, L)J \| U^{(\lambda)} \| (S', L')J' \rangle|^2 \qquad (3.22)$$

where the $U^{(\lambda)}$ terms are the matrix elements of unit tensor operators. It has been shown that the matrix elements of the majority of rare-earth ions are not very sensitive to the crystal-fields surrounding the lanthanide ion being considered, which in turn are functions of the electrostatic energy and spin-orbit parameters. The matrix elements $U^{(\lambda)}$ have been calculated and tabulated for transitions from the ground state by Carnell *et al.* [26]. The three intensity parameters Ω_2, Ω_4, and Ω_6, on the other hand, contain the effects of the crystal field terms but are independent of electric quantum numbers with the 4f configuration of the rare earth ions.

The line strength S can be related to the integrated absorbance of an electric dipole in the form of [19, 27]

$$\int_{\text{band}} k(\lambda) d\lambda = \rho\{8\pi^3 e^2 \lambda_a / 3ch(2J+1)\}(1/n)[(n^2+2)^2/9]S \qquad (3.23)$$

where $k(\lambda)$ is the absorption coefficient at wavelength λ, ρ is the ion concentration, λ_a is the mean wavelength of the absorption band, J is the total angular momentum of the initial level and $n = n(\lambda_a)$ is the bulk index of refraction at wavelength λ_a. Ω_λ parameters can therefore be determined as phenomenological parameters from a least squares fit of absorption bands.

Once these parameters for a given ion-glass combination have been derived, they can be used to calculate the strength of any absorption or emission transitions [19, 20]. The spontaneous emission probability is given in terms of the intensity parameters as

$$A(J, J') = \{64\pi^4 e^2/3h(2J+1)\lambda_a^3\}\{n(n^2+2)^2/9\}S \qquad (3.24)$$

where n is the index of refraction of the matrix. The factor $(n^2+2)^2/9$ represents the local field correction for the ion in a dielectric medium. The peak-stimulated emission cross-section, the radiative lifetime and the fluorescence branching ratio from level J to J', can be expressed as

$$\sigma_p = (\lambda_p^4/8\pi cn^2\Delta\lambda_{eff})A(J, J')$$
$$= \{\pi^3 e^3/3hc(2J+1)\}\{(n^2+2)^2/n\}(\lambda_p/\Delta\lambda_{eff})S \qquad (3.25)$$

$$\tau_R^{-1} = \sum_{J'} A(J, J') \qquad (3.26)$$

$$\beta_{JJ'} = A(J, J') \Big/ \sum_{J'} A(J, J') \qquad (3.27)$$

where λ_p is the peak wavelength of emission, $\Delta\lambda_{eff}$ is the effective line width and the summation is over all terminal level J'. The quantum efficiency η is given by

$$\eta = \tau_f/\tau_R. \qquad (3.28)$$

Here, τ_f is the measured fluorescence lifetime of the $|(S, L)J\rangle$ manifold.

3.1.7 Nonradiative relaxation [4, 13, 15, 17]

An excited rare-earth ion produced by absorption of a photon, also relaxes to the lower energy state via a nonradiative process. The population inversion density ΔN is proportional to the fluorescence lifetime τ_f, the number of atoms in the ground state N_0 and the pumping efficiency p_e:

$$\Delta N \sim N_0 p_e \tau_f \qquad (3.29)$$

The fluorescence lifetime is given by the radiative transition probability A_{rad} and the nonradiative transition probability W_{nr} in the Eq. (3.4). The radiative quantum efficiency and thus the lifetime of the initial lasing state will be lowered by nonradiative decay to a lower energy state. Therefore, to increase ΔN while maintaining a high stimulated emission cross-section σ, it is necessary to minimize the nonradiative transitions.

The decay rate W_{nr} is given by

$$W_{nr} = W_{mp} + W_{et} \qquad (3.30)$$

where W_{mp} and W_{et} are the multiphonon relaxation rate and the energy transfer rate, respectively. These processes are sensitive to the surrounding matrix ions. The general description and much useful information on these nonradiative transitions of rare-earth ions in glass has been covered recently by L. A. Riseberg and M. J. Weber [4], by R. Reisfeld [13, 15] and by K. Patek [17] in their excellent books.

3.1.7.1 *Multiphonon relaxation [4, 13, 15]*

As shown in Fig. 3.6.(a), multiphonon relaxation between J energy states can occur by the simultaneous emission of several phonons that are sufficient to conserve the energy of the transition. These processes arise from the interaction of the rare-earth ion with the fluctuating crystalline electric field due to the vibrations of the lattice or molecular groups, the excitation energies of which are determined by the masses of the constituent ions and the binding forces. When the energy gap between the excited level and the next lower energy level is larger than the phonon energy, several lattice phonons are emitted in order to bridge the energy gap. If ΔE is the energy gap between two energy levels and a number p of phonons with energy $\hbar\omega_{ph}$ is required for conservation of energy, the order of the process is determined by the condition

$$p\hbar\omega_{ph} = \Delta E. \qquad (3.31)$$

The dependence of the multiphonon rates on the energy gap is given phenomenologically in the form of [28, 29]

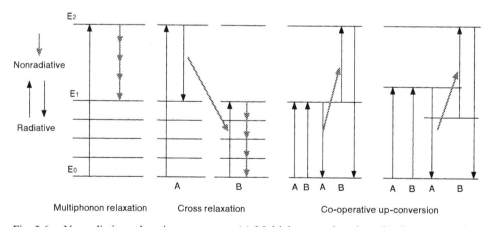

Fig. 3.6. Nonradiative relaxation processes, (a) Multiphonon relaxation, (b) Cross relaxation, (c) Co-operative up-conversion.

$$W_{mp} = W_0 \exp[-\alpha \Delta E] \tag{3.32}$$

where W_0 is the transition probability extrapolated to zero energy gap and α correlates with the phonon vibration ω_{ph} by [29]

$$\alpha = (1/\hbar\omega_{ph})\{\ln[p/g(n+1)] - 1\} \tag{3.33}$$

where g is the electron phonon coupling constant. This equation is an important result of the multiphonon theory because it shows the exponential dependence of W_{mp} on ΔE as well as the phonon vibration ω_{ph}.

3.1.7.2 Co-operative relaxation [4, 13, 17]

The remaining dissipative process to be considered involves ion–ion interactions primarily leading to concentration quenching, wherein the relative intensity of fluorescence is decreased as the concentration of the active ions is increased. Rare-earth ions in solids can be treated as isolated ions only when they are well separated. As the concentration is increased, the ion spacing may become sufficiently small that the ions interact with each other. Energy transfer may take place, where the original ion loses energy $(E_2 - E_1)$ nonradiatively by going down to the lower energy state E_1 and another ion acquires the energy by going up to a higher energy state. This phenomenon is called co-operative relaxation and occurs also in clusters, i.e. local groups of a few ions very close to each other. The coupling of adjacent rare-earth ions can arise via two specific relaxation processes schematically illustrated in Fig. 3.6.

The first process shown in Fig. 3.6(b) is a cross relaxation. Cross relaxation may take place between the same rare-earth ions, which happen to have two pairs of energy levels separated by the same amount. In this case, an excited ion A gives half its energy to a ground-state ion B, so that both ions end up in the intermediate energy level. From this level, they both relax quickly to the ground state via phonon relaxation.

Cross relaxation has been measured in a variety of ions and it is believed to be a dominating factor in the concentration quenching of Nd^{3+} ion in glass, which is associated with the reaction between $Nd^{3+}(^4F_{3/2})$–$Nd^{3+}(^4I_{15/2})$ and $Nd^{3+}(^4I_{9/2})$–$Nd^{3+}(^4I_{15/2})$ [13]. If an ion excited into the metastable $^4F_{3/2}$ level interacts with a nearby ion in the ground state, the first ion can transfer a part of its energy to the second, leaving both in the intermediate $^4I_{15/2}$ state. Since the energy gap to the lower-lying states are small, both ions quickly decay nonradiatively to the ground state.

The probability for dipole-dipole transitions of two ions A and B separated by distance R_{AB} is [17, 30, 31]

$$P_{AB} = K(1/R_{AB})^6(\beta/\tau_R)S_A \int [f_A(\nu)g_B(\nu)\nu^{-4}\,d\nu] \qquad (3.34)$$

where K is a constant, R_{AB} is A–B ionic distance, β is the branching ratio of the fluorescence band, τ_R is the radiative lifetime of the ion A, S_A is absorption band strength, $f_A(\nu)$ and $g_B(\nu)$ are the emission and absorption band spectral line shapes, respectively, and ν is frequency. The host dependence arises from the different range and distribution of ion–ion separations R_{AB}, and the different frequency and linewidths of the $^4F_{3/2}$–$^4I_{15/2}$: $^4I_{9/2}$–$^4I_{15/2}$ transitions.

The second one, shown in Fig. 3.6(c), is a co-operative up-conversion process [32, 33]. As in an Auger process, when two neighboring ions are in the excited state, one of them (A) gives its energy to the other (B), thus promoting B to a higher energy level while the A relaxes to the ground state. From this higher energy level, the B ion relaxes rapidly, mostly down to the metastable state via phonon relaxation. The major cause of concentration quenching for Er^{3+}-doped glasses is believed to be due to this process [32, 33]. Er^{3+} has no intermediate states between the metastable and ground state, thus cross relaxation between an excited ion and one in the ground state cannot occur. However, if two excited ions interact, one can transfer its energy to the other, leaving itself in the ground state and the other in the higher $^4I_{9/2}$ state.

3.2 Bulk laser glasses

3.2.1 Compositional dependence of laser parameters of Nd^{3+}-doped glasses

Since the first demonstration of laser action in Nd^{3+}-doped glass by Snitzer [34], many rare-earth ions have displayed laser transitions in near-ultraviolet, visible and near-infrared spectral regions. In addition, there are numerous possibilities of incorporating rare-earth ions into the host glasses. Good views and textbooks related to this subject have been provided by many authors [8, 10, 13, 15–17]. Over the past two decades, however, the development of bulk glass lasers have concentrated primarily on Nd^{3+}-doped laser glass. Its high efficiency is thought to be ideal for use in the high-peak-power lasers needed, for example, for inertial confinement nuclear fusion research [35].

The design of a high-peak-power glass laser system involves consideration of a number of spectroscopic parameters [35]. For example, for continuous-wave (CW) operation the threshold for laser oscillation is inversely proportional to the product of the lifetime τ and emission cross section σ_e. Since rare-earth ions can be incorporated as modifier ions into most glasses, the variability in the sizes and valence states of known glass-forming and glass-modifying cations and anions produces a very large range of possible local fields at an

Table 3.1. *Sets of reduced matrix elements of transitions from the $^4F_{3/2}$ level of Nd^{3+}-ion [19]*

Transition	Wavelength (mm)	Reduced matrix elements, $U(\lambda)$		
		$U(2)$	$U(4)$	$U(6)$
$^4F_{3/2} \rightarrow {}^4I_{9/2}$	0.88	0.0	0.230	0.056
$^4F_{3/2} \rightarrow {}^4I_{11/2}$	1.06	0.0	0.142	0.407
$^4F_{3/2} \rightarrow {}^4I_{13/2}$	1.35	0.0	0.0	0.212
$^4F_{3/2} \rightarrow {}^4I_{15/2}$	1.88	0.0	0.0	0.0285

Note: $U(\lambda) \equiv |\langle(^4F_{3/2})\| U^{(\lambda)} \|(S', L')J'\rangle|^2$.

active ion site. To achieve the maximum performance of the laser system and also for an understanding of the interaction of rare-earth ions with the local environment in glass, the spectroscopic properties of Nd^{3+} ion in various glass compositions have been studied extensively during the past 30 years. Good reviews and textbooks related to this subject have been provided by D. C. Brown [35], M. J. Weber [10], R. Reisfeld [13, 15], and M. J. Weber and J. E. Lynch [36].

The method for predicting the value of spectroscopic parameters for a given composition has been developed by the application of the Judd–Ofelt (J–O) approach [23, 24] which was briefly summarized in Section 3.1.6. It generally requires precise measurements of fluorescence intensity to determine spectro-scopic parameters experimentally. Inhomogeneity and the size of the samples available, however, limits such measurements. By using the J–O theory, it generally only requires measurements of absorption spectra to determine the best set of intensity parameters. Data obtained by examination of small laboratory samples of glass compositions can be used to predict the spectro-scopic parameters accurately [4, 35].

The first report of the application of the J–O approach to Nd^{3+} in glass was by W. F. Krupke [19], who calculated emission cross sections, radiation lifetimes, and branching ratios for four commercial laser glasses. He has shown that the J–O theory can be applied to Nd^{3+} laser glass and can successfully account for the laser parameters. The line strength S was at first calculated from Eq. (3.23). A computerized least-squares fitting procedure was then employed that determines the values of the phenomenological parameters Ω_λ by use of Eq. (3.28) and the reduced matrix elements of unit-tensor operators (see Table 3.1.). With the determination of Ω_2, Ω_4, and Ω_6, it is a simple matter to determine the radiative probability, branch ratios, radiative lifetime, quantum efficiency and peak induced-emission cross section respectively.

Table 3.2. Range of intensity parameters of transitions from the $^4F_{3/2}$ level of Nd^{3+} ion in various oxide glasses [Reprinted from R. Jacobs and M. Weber, IEEE J. Quantum Electron. **QE-12** (1976) 102, copyright (1976) with permission from the Institute of Electrical and Electronics Engineeers, Inc.]

| Compositional change | Range of intensity parameters Ω_λ (10^{-20} cm²) | | | $^4F_{3/2}$–$^4I_{11/2}$ transition | | | Radiative lifetime |
	Ω_2	Ω_4	Ω_6	Cross section σ_p (pm²)	Wavelength λ_p (nm)	Linewidth $\Delta\lambda_{eff}$ (nm)	τ_R (μs)
Glass former	1.6 ~ 6.0	3.2 ~ 5.4	2.6 ~ 5.4	1.5 ~ 4.1	1054 ~ 1069	25.3 ~ 43.1	239 ~ 624
Alkali in silicate glass	3.3 ~ 5.0	2.4 ~ 4.9	2.0 ~ 4.5	0.97 ~ 2.6	1058 ~ 1061	33.4 ~ 37.7	396 ~ 942
Alkali in alkaline earth silicate	2.8 ~ 4.2	2.9 ~ 4.0	2.5 ~ 3.9	1.4 ~ 2.3	1061 ~ 1062	33.6 ~ 36.0	412 ~ 657
Alkaline earth in alkali silicate	3.7 ~ 5.1	2.8 ~ 4.1	2.3 ~ 2.8	1.4 ~ 1.7	1061	35.1 ~ 41.2	547 ~ 657
Silica-sodium in alkali silicate	4.0 ~ 6.1	2.9 ~ 4.5	2.6 ~ 4.3	1.5 ~ 1.9	1059 ~ 1088	33.0 ~ 44.8	483 ~ 694

R. R. Jacobs and M. J. Weber [20, 25] have also utilized the J–O approach to evaluate the radiative parameter of Nd^{3+}: $(^4F_{3/2})$–$(^4I_{11/2})$ transition for a number of silicate and phosphate glasses, which was reviewed later by D. C. Brown [35]. Their results, which list $\Delta\lambda_{eff}$, λ_p, τ_R, and σ_p for the $(^4F_{3/2})$–$(^4I_{11/2})$ transition, are summarized in Table 3.2, together with the variation of intensity parameters Ω_λ. These results also show clearly that changes in intensity parameters occur when the glass composition is varied. They studied a series of five variations of composition and discussed the changes of $\Delta\lambda_{eff}$, λ_p, β, τ_R, and σ_p. The first series involved changes of only the glass network former in eight different glasses, namely phosphate, borate, germanate, silicate, tellurite, aluminate, titanate and fluorophosphate. The branching ratio for all transitions was relatively invariant with composition, whereas the total transition rate (or lifetime) varied significantly. For the $(^4F_{3/2})$–$(^4I_{11/2})$ transition, λ_p varied from 1055 nm in the phosphate to 1069 nm in the aluminate glass. The cross section σ_p is a maximum of 4.1×10^{-20} cm^2 in the phosphate glass and a minimum of 1.5×10^{-20} cm^2 in the silicate. Another important fact to emerge from this work is that, as the composition is changed, increasing the M^+ radius in the sequence Li–Na–K–Rb caused σ_p to decrease and τ_R to increase.

The J–O approach was also applied to nine commercially available laser glasses [20, 25]. The phosphate glasses displayed large cross sections and a shorter center wavelength than silicates. This is not only because the intensity parameters were significantly larger but also because the line widths were narrow. From Eq. (3.25) it should be noted that the peak cross section σ_p is dependent upon Ω_λ parameters as well as $\Delta\lambda_{eff}$, and that both parameters are very composition dependent. The emission bandwidths of Nd^{3+}-doped phosphate glasses are typically narrower than those for silicates. The combination of large Ω_λ values and narrow line widths can yield larger peak cross sections for the phosphate glasses [10, 37].

Among different glass-forming systems, the narrowest line widths occur when the rare earth can establish the most symmetric, chemically uniform coordination sphere with the weakest field. The narrow line widths of phosphate glasses are thought to be a consequence of their different structure; a chain-like structure versus the random network structure of silicate glasses [10, 37].

A large amount of useful information has been published by the Lawrence Livermore Laboratory for a large number of commercially and scientifically important laser glasses [36]. Table 3.3 summarizes the ranges of effective stimulated-emission cross sections, σ_p, for the $(^4F_{3/2})$–$(^4I_{11/2})$ transition of Nd^{3+} observed for different glass types; the associated ranges for n, $\Delta\lambda_{eff}$, λ_p, and the calculated radiative lifetime, τ_R are also given. For silicate glasses, σ_p

Table 3.3. *Range of spectroscopic properties for the $^4F_{3/2}-^4I_{11/2}$ transition of Nd^{3+} observed in different glasses at 295 K [Reprinted from M. J. Weber, J. Non-Cryst. Solids **123** (1990) 208, copyright (1990) with permission from Elsevier Science]*

Host glass	Refractive index n_d	Cross section σ_p (pm^2)	Peak wavelength λ_p (µm)	Effective linewidth $\Delta\lambda_{eff}$ (nm)	Radiative lifetime τ_R (µs)
Oxide					
silicate	1.46–1.75	0.9–3.6	1.057–1.088	35–55	170–1090
germanate	1.61–1.71	1.7–2.5	1.060–1.063	36–43	300–460
tellurite	2.0–2.1	3.0–5.1	1.056–1.063	26–31	140–240
phosphate	1.49–1.63	2.0–4.8	1.052–1.057	22–35	280–530
borate	1.51–1.69	2.1–3.2	1.054–1.062	34–38	270–450
Halides					
beryllium fluoride	1.28–1.38	1.6–4.0	1.046–1.050	19–29	460–1030
aluminium fluoride	1.39–1.49	2.2–2.9	1.049–1.050	28–32	540–650
heavy metal fluoride	1.50–1.56	2.5–3.4	1.048–1.051	25–29	360–500
chloride	1.67–2.06	6.0–6.3	1.062–1.064	19–20	180–220
Oxyhalides					
fluorophosphate	1.41–1.56	2.2–4.3	1.049–1.056	27–34	310–570
chlorophosphate	1.51–1.55	5.2–5.4	1.055	22–33	290–300
Chalcogenides					
Sulfide	2.1–2.5	6.9–8.2	1.075–1.077	21	64–100
oxysulfide	2.4	4.2	1.075	27	92

varies by a factor of four with changing host composition from an alkali silicate to bismuth silicate, which has a high refractive index. The spontaneous emission probability A depends upon the host refractive index, is a function of $[(n^2+2)^2/9]$, and thus the peak cross section σ_p is a function of $[(n^2+2)^2/9n]$. These factors arise from the local field correction, and can have a significant effect upon the values of A and σ_p. Some tellurite glasses have even larger cross sections than phosphates, not because S or $\Delta\lambda_{eff}$ are more favorable, but because of their large refractive indices.

According to the reviews of M. J. Weber *et al.* [10, 36], the smaller the anionic field strength, in general, the smaller the Stark splitting and the narrower the effective line widths. Thus $\Delta\lambda_{eff}$ is narrowed by using monovalent halide rather than divalent oxide anions. As expected, the halide glasses have narrower line widths than the narrowest oxide glass (phosphates). Chalcogenide glasses combine a small $\Delta\lambda_{eff}$ with a larger Ω_λ and a high refractive index to yield values for σ_p that are larger than those for any oxide glass. The use of larger, more polarizable anions, such as Cl⁻ or Br⁻ instead of F⁻, results in both narrow linewidth and increased intensity parameters. Chloride glasses have larger cross sections because of their larger refractive index.

The meaning of intensity parameters on the basis of the relationship between the glass structure and Ω_λ for a given rare-earth ion in various glass systems has also been discussed by many researchers [13–15, 38–42].

3.2.2 Nonradiative relaxation of rare-earth ions in glasses

The dependence of the multiphonon rate of energy gap has been investigated by C. B. Layne *et al.* [43, 44]. The rare earths Nd^{3+}, Er^{3+}, Pr^{3+}, and Tm^{3+} were doped into silicate laser glass and the multiphonon rates measured experimentally; the results are shown in Fig. 3.7. As can be seen from Eq. (3.32), an approximately exponential dependence upon the energy gap between levels is found, independent of the rare-earth ion or electronic level.

Because the multiphonon process is dominated by the highest energy phonons which are known to vary widely in different glass compositions,

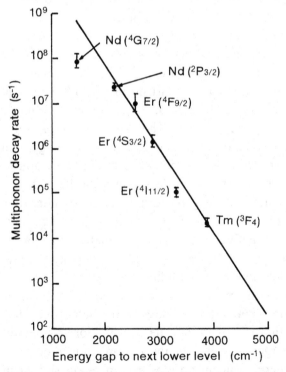

Fig. 3.7.   Mulitphonon decay rates as a function of energy gap to the next lower level for rare earths in silicate glasses. [Reprinted from C. B. Layne, W. H. Lowdermilk and M. J. Weber, *Phys. Rev.* **B16** (1977) 10, copyright (1977) with permission from The American Physical Society.]

Table 3.4. *Maximum phonon energy of various glasses [13]*

Glass type	Phonon energy $\hbar\omega_{ph}$ (cm^{-1})
Borate	1400
Silicate	1100
Phosphate	1200
Germanate	900
Tellurite	700
Fluorozirconate	500
Chalcogenide	350

significant variations in the relaxation rate would be expected with glass type. The glasses most favorable for use in lasers tend to have the lowest phonon $\hbar\omega_{ph}$ which provides smaller multiphonon relaxation rates W_{mp}. In any glass, the most energetic vibrations are the stretching vibrations of the glass network formers, which are active in the multiphonon process [13]. R. Reisfeld and C. K. Jørgensen [13] have listed the phonon energies of importance in various types of glass and they are reproduced in Table 3.4.

The results obtained in a study of multiphonon rates in borate, phosphate, silicate, germanate, tellurite and fluoroberyllate glasses by C. B. Layne *et al.* [43–45], in fluorozirconate by Lucas *et al.* [46], and in zinc-chloride glass by M. Shojiya *et al.* [47] are also summarized in Fig. 3.8 which shows the same approximate exponential dependence of multiphonon rate on energy gap as in Fig. 3.7, for each glass composition. The fastest decay rates are found in the borate glass, followed in order by phosphate, silicate, germanate, tellurite, fluorozirconate, fluoroberyllate, and chloride. The multiphonon decay rate from a given level to the next lower level decreases with the lowering of energy of the stretching frequencies of the glass former. Since, in order to reach the same energy gap, a larger number of phonons is needed in fluoride and in chalcogenide glasses, the nonradiative relaxations are smallest in these glasses. Additional information on multiphonon relaxation in glasses may be found in Section 3.3.3 which is concerned with fiber laser amplifiers.

When OH$^-$ groups are present in a glass, the nonradiative rates are increased because of coupling between the rare earth states and the high energy (3200 cm^{-1}) vibrations of OH$^-$ [48]. Its contribution to nonradiative decay is considered to be larger than that of the multiphonon process because of the higher vibration energy. Phosphate glasses can retain significant amounts of OH$^-$ and consequently the lifetimes of the rare-earth states are shortened. The water content must be controlled to be as low as possible.

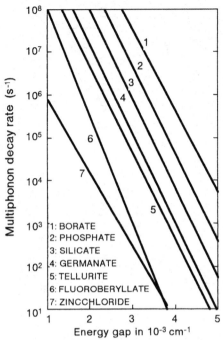

Fig. 3.8. Multiphonon decay rate of rare-earth ions in several host glasses as a function of the energy gap to the next lower level. [Oxide glasses, reprinted from C. B. Layne, W. H. Lowdermilk and M. J. Weber, *Phys. Rev.* **B16** (1977) 10, copyright (1977) with permission from The American Physical Society. Fluorozirconate glass, reprinted from J. Lucas and J-L. Adam, *Glastech. Ber.* **62** (1989) 422, copyright (1989) with permission from Publishing Company of the German Society of Glass Technology. Zinc-chloride glass, reprinted from M. Shojiya, M. Takahashi, R. Kanno, Y. Kawamoto and K. Kadono, *Appl. Phys. Lett.* **72** (1998) 882, copyright (1998) with permission from the American Institute of Physics.]

Interactions between rare-earth ion pairs also lead to nonradiative decay or concentration quenching, of excited states. The severity of concentration quenching, however, depends on the glass type and composition and it is least in phosphate glasses. In fact, a glass that is doped with 8 wt% of Nd^{3+} ions has been developed by using phosphate glass [49], which is probably attributed to the large Nd–Nd distance resulting from the chain-form structure, and to the difficulty of Nd–Nd interaction [31].

3.2.3 Properties of practical laser glasses

As described above, large variations in stimulated emission cross section and other important radiative parameters are realized by varying the host composition. However, depending on the applications, other properties in addition to the spectroscopic properties are required for practical laser glasses.

Table 3.5 summarizes the spectroscopic, optical, thermo-optic and mechani-

Table 3.5. *Lasing parameters of commercial laser glasses [50]*

	High gain silicate LSG91H	High gain and low n_2 phosphate LHG5	Athermal phosphate LGH8	Super-strong high power HAP4	Low n_2 fluorophosphate LHG10	Er^{3+} doped eye safe LEG30
Lasing Properties						
Nd^{3+} conc. (10^{20}/cm^3)	3.0	3.2	3.1	3.2	3.0	—
Stimulated emission cross section σ_p (10^{-20} cm^2)	2.7	4.1	4.2	3.6	2.7	7.7
Fluorescence lifetime τ_f (μs)	300	290	315	350	380	8700
Fluorescence halfwidth $\Delta\lambda$ (A)	274	220	218	270	265	320
Center lasing wavelength, λ_p (mm)	1062	1054	1054	1054	1054	1535
Optical Properties						
Refractive Index, n_d	1.56115	1.54096	1.52962	1.5433	1.4592	1.54191
Abbe number, ν_d	56.6	63.5	66.5	64.6	89.9	65.4
Temperature coefficient of refractive index, $(dn/dT)(10^{-6}/°C)(20°C-40°C)$	+1.6	−0.4	−5.3	+1.8	—	−3.0
Temperature coefficient of optical path length $(ds/dT)(10^{-6}/°C)(20°C-40°C)$	+6.6	+4.2	+0.6	+5.7	—	+2.2
Nonlinear refractive index, $n_2(10^{-13}$ esu$)$	1.58	1.28	1.13	1.25	0.61	1.22
Other properties						
Thermal conductivity (25 °C) (kcal/m, h, °C)	0.89	0.66	0.5	0.88	0.74	0.47
Coefficient of linear thermal expansion $(10^{-7}/°C)(20°C-40°C)$	90	84	112	72	153	96
Fracture toughness (10^{-3} MN/m$^{3/2}$)	—	540	—	830	—	—
Thermal shock resistance (kJ/m$^{1/2}$h)	—	2.23	—	3.67	—	—

cal properties of the examples of commercially available Nd-laser glasses [50]. These glasses are representative of laser glasses that can be melted in large sizes with high optical quality.

The laser glass is strongly loaded by pump radiation and by the laser beam itself. In comparison with crystals, the major disadvantage of glass is its low thermal conductivity. Therefore, the temperature distribution in the laser glass produced by this load causes optical distortion which is affected by the change of refractive index with the temperature, which limits its applicability in high-average-power systems. The temperature coefficient of the optical path length is given by [51–53]

$$ds/dT = [dn/dT + (n-1)\alpha] \qquad (3.35)$$

where n is the refractive index and α is the thermal expansion coefficient. Thus, the chemical composition of the laser glasses should be selected so that ds/dT is minimized at a given operating condition (athermality). Phosphate glass has a fairly high σ_p value and small linewidth, which is good for large signal amplification. In addition, the temperature coefficient of the optical path length, ds/dT, is very low for phosphate glasses.

Another measure of the suitability of a particular laser glass for an application in which thermal cycling is involved is the thermal-shock resistance. Thermomechanical properties of laser glasses are critically important in high-average-power lasers, in which the heat removal rates and thermal stresses determine the maximum power. The thermomechanical figure of merit, W_{tm}, at its fracture limit is given by [54, 55]

$$W_{tm} = \{\kappa(1-\mu)/\alpha E\}S_t \qquad (3.36)$$

where E is the Young's modulus, α is the thermal expansion coefficient, κ is the thermal conductivity, μ is Poisson's ratio, S_t is fracture strength. For future increases in the power and repetition rate of high average power glass lasers, new silico-phosphate glasses (for instance, HAP 3 in Table 3.5) that might improve thermo-mechanical properties and longer fluorescence lifetimes, have been developed [56].

High-peak-power glass laser systems commonly propagate high optical intensities. Such intensities give rise to an intensity-dependent change to the local refractive index through the third-order nonlinear susceptibility of glass, which will be discussed in detail in Chapter 4. The existence of a nonlinear index change is responsible for the well-known self-focusing effect. Fluorophosphate glasses appear to be good candidates for high-power lasers because of their low nonlinear index change [57] but the fabrication of these glasses in large quantities is too expensive.

3.3 Fiber lasers and amplifiers

3.3.1 General description of fiber laser

3.3.1.1 History

Fiber lasers are nearly as old as the glass laser itself. Three years after the first demonstration of a glass laser by E. Snitzer in 1961 [34], C. J. Koester and E. Snitzer [58] described the first lasing in a multicomponent glass fiber by demonstrating amplification. A decade later, Stone and Burrus [59, 60] described the laser oscillation of Nd-doped low-loss silica multimode fibers made using a CVD process as well as the first diode pumping. D. N. Payne and co-workers at Southampton University [61–63] announced the fabrication of the first low-loss single mode fibers containing Nd and Er, made by the extension of MCVD technology and reawakened interest in fiber lasers, which coincided with its large-scale application to optical communication systems. Fiber lasers have potential for use in both long-haul (1.5 μm) and local area network (1.3 μm) communications as optical amplifiers. Noncommunication applications include use as super luminescent sources for fiber-optic gyroscopes and low-cost sources in various instruments for optical measurements. P. Urquhart [64] carried out an early review of rare-earth doped fiber laser amplifiers. The many recent advances in the development of rare-earth doped fiber lasers and amplifiers is also covered by other reviews [8, 65–67].

3.3.1.2 Benefits of lasers in fiber form

Optical fibers are basically waveguides [68, 69] for electromagnetic radiation at optical frequencies. When light passing through the fiber core is incident at the interface between the core and cladding which have different refractive indices (n_1 and n_2 respectively), it will be totally reflected if its angle of incidence is greater than the critical angle ($\sin \phi_c = n_2/n_1$). The numerical aperture (*NA*) of the fiber, which is a measure of the light-gathering power is given by

$$NA = (n_1^2 - n_2^2)^{1/2}. \tag{3.37}$$

Other important properties of a fiber are associated with guided modes, which are primarily governed by the waveguide structure and materials. On this basis, fibers can be classified into two main types: monomode and multimode. The simplest type is the monomode fiber, in which the core size and refractive index difference are small, allowing only the waveguide mode of lowest order to propagate. The condition necessary for this to occur is that the normalized frequency (*V* number) defined by

$$V = (2\pi a_c/\lambda)(n_1^2 - n_2^2)^{1/2} \tag{3.38}$$

should be less than 2.405: here a_c is the core radius and λ the operating wavelength. The increase in core size allows many modes to propagate. Multimode fibers with much larger cores have the numbers of modes ($\sim V^2/2$) that is typically about 1000.

One of the benefits of lasers in fiber form is that a large population inversion density is attainable, as the result of the guiding structures of the fiber by keeping the pump light tightly confined in the transverse direction over the distance required for its absorption [8, 64]. The high surface-area-to-volume ratio, inherent in optical fiber geometry, leads to a good heat dissipation. These factors, taken together, explain why fiber lasers with glass hosts readily operate in continuous wave mode at room temperature without cooling, whereas bulk glass lasers with a low thermal conductivity can usually only operate to give a pulsed output. A theoretical analysis of optically pumped fiber laser amplifiers and oscillators [70] suggests that fiber configuration will have higher gain (amplifiers) and lower threshold (oscillators) over its bulk counterpart.

3.3.1.3 Fabrication techniques for fiber lasers

The fiber fabrication process, in general, consists of two steps, the preparation of the bulk preform and the formation of a long fiber. There are many variations in the preform preparation and fiber formation processes, depending on the fiber materials. Details on the process of fiber fabrication without rare-earth dopant can be found in Chapter 1 of this book. For more information concerning the fabrication process, there are excellent reviews for silica fibers by P. Urquhart [64], W. A. Gambling [71], K. J. Beales and C. R. Day [69] and a book by T. Izawa and S. Sudo [72]; for fluoride fibers by H. Poignant *et al.* [73] and by T. Miyashita and T. Manabe [74]; and for chalcogenide fibers by J. Nishii *et al.* [75].

The high silica glass systems are most preferable for fiber laser host materials because they have low-loss transmission windows at about 1.3 μm and 1.55 μm which are used in the optical communication systems. The rare-earth-doped silica fiber preform is in general fabricated using the CVD technique, with a number of important modifications to permit the incorporation of further dopants into the core glass. During the core deposition, small quantities of $NdCl_3$ vapour are carried downstream by the reactant flow, where they are oxidized to Nd_2O_3 in the hot zone formed by the deposition burner and incorporated into the core [61, 63]. Dopant can also be incoporated into the fiber by soaking the porous preform in a solution of a chloride of the rare-earth ion in alcohol or in an aqueous solution of the required dopant precursor, prior to dehydrating, sintering, sleeving and collapsing (solution doping technique). Control of dopant levels is easily achieved and high dopant levels have been obtained using this method [59, 67, 76, 77].

Fluoride glasses with low phonon energy, which were originally developed for ultra-low-loss communication systems [64, 74], make extremely good hosts for rare-earth ions because they offer many more metastable fluorescing. Heavy-metal fluoride glasses are often unstable when cooled and show a tendency to crystallize in the fiber drawing stage. Glass preforms with a core and cladding structure have been prepared by special techniques, such as built-in casting [78, 79], suction casting [80, 81], and rotational casting [82]. In order to prevent the crystallization and vaporization, a quite unique technique, namely the 'extrusion method' [83] as shown in Fig. 3.9, is used for a fluoride preform fabrication. Both core and clad glass discs were polished and placed inside the cylinder. Then, the system was heated up followed by pressuring with a punch under a dry nitrogen atmosphere. In this method, a preform can be formed at higher viscosity, and hence at a lower temperature compared with the conventional casting methods, which prevents crystallization.

The chalcogenide fiber is usually prepared using the double crucible and rod-in-tube methods [73]. New crucible drawing methods have been developed recently [75].

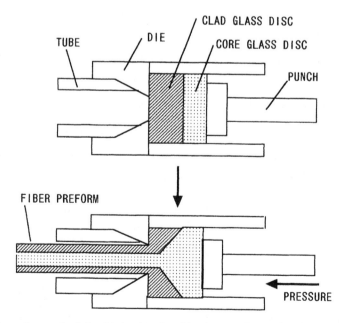

Fig. 3.9.   Extrusion method for fabricating fluoride glass preform. [Reprinted from K. Itoh, K. Miura, I. Masuda, M. Iwakura and T. Yamashita, *J. Non-Cryst. Solids* **167** (1994) 112, copyright (1994) with permission from Elsevier Science.]

3.3.2 Fiber amplifiers

3.3.2.1 General descriptions [8, 64, 67]

A driving force for research in rare-earth-doped fibers is their potential use for amplifying weak signals for trunk optical communication systems at 1.55 μm, and for both local subscriber systems in the near future and most of the installed optical systems at around 1.3 μm. The amplification of optical signals in a communication system may be achieved by simply splicing a section of rare-earth-doped fiber into the transmission fiber and injecting pump light through a fiber coupler. The signal within the emission band stimulates radiation, amplifying the signal with high gain, high efficiency and low noise, which is advantageous to optical communication [8, 64, 67]. A typical experimental configuration of a fiber-based optical amplifier [84] is shown in Fig. 3.10.

There are four principal rare-earth candidates for use as a dopant for fiber amplifiers in optical communication systems: erbium, neodymium, praseodymium and dysposium. Figure 3.11 shows a simplified energy diagram of these ions. The $^4I_{15/2}$–$^4I_{13/2}$ transition of Er^{3+} has been identified for the 1.55 μm window. The $^4F_{3/2}$–$^4I_{13/2}$ transition of Nd^{3+}, the $^1G_4$–$^3H_5$ transition of Pr^{3+} and the $^6F_{11/2}(^6H_{9/2})$–$^6H_{15/2}$ transition of Dy^{3+} have been identified as potentially useful for the 1.3 μm window. These ions, however, have their own set of difficulties that must be overcome in order to use them in practical applications [8, 32, 65–67]. The major problems are related to the dissipative process which removes excited ions from the initial lasing level of interest.

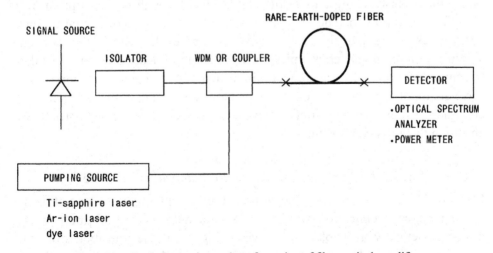

Fig. 3.10. Typical experimental configuration of fiber optical amplifier.

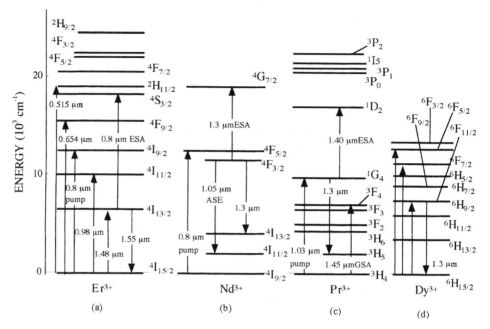

Fig. 3.11. Simplified energy diagrams of four rare-earth ions and relevant transitions for pumping and amplification. (Energy levels are from [18].)

One possible mechanism is nonradiative relaxation directly to the next lower energy level as treated in Section 3.1.7. The other dissipative processes which need to be considered involve the phenomena of excited-state absorption (ESA) and amplified spontaneous emission (ASE). ESA is the excitation of an ion in the metastable state to a still higher level through the absorption of a pump or signal photon. The ions promoted to a higher state of excitation then decay back nonradiatively to the upper lasing level. ASE is the efficient amplification of spontaneous emission from the upper laser level via the other transition with a high branching ratio, which causes gain saturation at the wavelength of interest. All of these phenomena decrease the efficiency of amplifiers, but are sensitive to host compositions and can be solved partially by changing the glass compositions.

3.3.2.2 1.55 μm amplification in Er^{3+}-doped fiber [32, 65]

Er^{3+}-doped fiber amplifiers have attracted much attention for operation in the optical communication window at wavelengths near 1.55 μm and are now commerically available. As shown in Fig. 3.11, the transition from the $^4I_{13/2}$ metastable state to the $^4I_{15/2}$ ground state produces a gain at 1.55 μm.

The 6500 cm^{-1} gap between two states is large enough that multiphonon relaxation is significant only for borate glasses with high phonon energy [32, 65].

Most work on Er^{3+}-doped amplifiers has focused on silica fiber. Silica would be the preferred glass because of its superior optical, mechanical and chemical properties coupled with mature manufacturing techniques, and its inherent compatability with the passive transport fiber. Pure silica, however, can incorporate only very small amounts of rare-earth ions. The absence of a sufficient number of non-bridging oxygens to coordinate the isolated rare-earth ions in the rigid silica network causes them to cluster in order to share nonbridging oxygens. Arai *et al.* [85] found that the addition of a small amount of Al_2O_3 can eliminate the clustering problem. Al_2O_3 forms a solvent shell around Er^{3+} ions, increasing the solubility and allowing dopant levels up to about $1 \times 10^{20} \text{ cm}^{-3}$. P-doping is also found to improve amplifier efficiency somewhat [85]. Furthermore, a low concentration of Er^{3+}, is more preferable because of the three-level laser characteristics in which efficient conditions at low pump power can only be obtained by using a small core and long distance fiber with low Er^{3+} concentration.

As shown in Fig. 3.11, Er^{3+} can be pumped either directly into the $^4I_{13/2}$ metastable level which occurs at 1490 nm or into higher energy levels, e.g. $^2H_{11/2}$ (514.5 nm), $^4F_{9/2}$ (654 nm), $^4I_{9/2}$ (800 nm) and $^4I_{11/2}$ (980 nm) from which nonradiative decay occurs to the $^4I_{13/2}$ metastable level. In early experiments, pumping at 514.5 nm with an Ar-ion laser or at 654 nm with a dye laser were employed [86]. Pumping at 800 nm has been examined recently in research into the best materials because of the availability of inexpensive high power AlGaAs diode lasers at this wavelength [87]. In most glasses, however, the performance of fiber pumped in this band has been rather poor because of the $^4I_{13/2}$–$^2H_{11/2}$ ESA transitions at 790 nm [32, 88].

The $^4I_{15/2}$–$^4I_{11/2}$ transition, corresponding to an absorption band peaking at $970 \sim 980$ nm, has a large absorption band cross section and no ESA was observed for this band. Thus 980 nm appears to be an ideal pump wavelength [88, 89] but inexpensive diode lasers are unavailable [65].

Since the $^4I_{13/2}$ Stark levels are nonuniformly populated, the $^4I_{15/2}$–$^4I_{13/2}$ transition is shifted to a longer wavelength with respect to absorption. This allows resonant pumping into the top of the band with 1480 nm light in a quasi three-level scheme [32, 65, 90–92]. The wavelength is close to the amplifier wavelength of $1.53 \sim 1.55$ μm and so there is minimum heating and rapid relaxation between sublevels of the metastable level, although a careful selection of pump wavelength is required.

Table 3.6. Amplification characteristics of rare-earth-doped fibers in the 1.3 μm region

Ion	Concentration (ppm, wt)	Host glass	Fiber specification Core diam. (μm)	Cut-off (nm)	Length (m)	Signal Wavelength (nm)	Pumping Wavelength (nm)	Pumping efficiency Max. gain/pump power, (dB)/(mW)	Efficiency (dB/mW)	Ref.
Nd	500	Fluorozirconate	5.8	1200	0.75	1320 ∼ 1337	795(TS)	4.5 ∼ 6.5/100		100
	<3wt%	Phosphate	10			1330 ∼ 1360	800(TS)	2.6/100		104
	2000	Fluorozirconate	6.5	1310	2	1319 ∼ 1339	820(TS)	3.3 ∼ 5.5/50	0.13	105
	1000	Fluorozirconate	6.5	1310	10	1330	496(Ar)	10/150		108
	500	Fluorozirconate	9		1.4	1320 ∼ 1350	795(TS)	4 ∼ 5/	0.03	101
	1 mol%	Fluorophosphate	7		0.42	1320	819(TS)	3.4/85		102
	2000	Fluorozirconate	6.5	1310	2.0	1319 ∼ 1343	820(TS)	7.1 ∼ 10/160	0.07	116
Pr	500	Fluorozirconate	4		8	1310	1017(TS)	7.1/180		111
	500	Fluorozirconate	3.3	650	23	1309	1017(TS)	32.6/925	0.04	106
	2000	Fluorozirconate	4	1180	2	1310	1017(TS)	13/557	0.022	107
	500	InF$_3$/GaF$_3$			10.2	1310	1017(TS)	26/135	0.25	110
	1200	Fluorozirconate	3.5		4	1319	1064(NY)	15		112
	560	Fluorozirconate			17	1320	1007(TS)	10.5/550	0.019	113
	500	Fluorozirconate	2		10	1310	1017(LD)	5.2/41	0.2	109
	500	Fluorozirconate	2	900	17	1310	1017(LD)	15.1/160	0.145	114
	1000	Fluorozirconate	–	850	13	1310	1013(MP)	26/380	0.11	115
	500	Fluorozirconate			40	1300	1017(LD)	28.3/		117
	2000	Fluorozirconate	2.3	1260	8	1310	1017	38.2/300	0.21	118
	1000	Fluoride	1.8	950	15	1302	1047(NYL)	20/550		119
	1000	Fluoride	1.8	950	14 & 22	1300	1047(NYL)	40.6/550		120
	1000	Fluoride	1.7	900	10	1300	1017(MP)	30.5/500		121
	500	PbF$_2$/InF$_3$	1.2	1000	19	1300	(MP)	28.2/238	0.36	122
	1000	InF$_3$/GaF$_3$	1.1		8.5	1311	1017(TS)	29/215	0.29	123
	1000	InF$_3$/GaF$_3$	2.8	800		1311	1004(TS)	39.5/100	0.65	124
	750	Ga–Na–S	2.5		6.1	1340	1017(TS)	30/100	0.81	125
	1000	PbF$_2$/InF$_3$			20	1305	1015(TS)		0.40	126
Pr/Yb	3000/2000	Fluorozirconate	1.7		1		1017(LD)	3.2/75	0.055	109
	2000/4000	ZBLAN	4.0	960	2	1310	980	∼6/5		118

TS: Ti:Sapphire, LD: Laser diode, Ar: Ar ion laser, NY: Nd:YAG, NYL: Nd:YLF, MP: Master oscillator/Power Amplifier Laser Diode.

3.3.2.3 1.3 μm amplification in Nd^{3+}-doped fiber

Nd^{3+} was the first ion considered for amplifiers in the 1.3 μm window because it has a well-known four-level $^4F_{3/2}$–$^4I_{13/2}$ transition close to 1.3 μm as shown in Fig. 3.11. Furthermore this system can be pumped by a commercially available, high power diode laser (AlGaAs) source in the 800 nm region. The energy difference between the $^4F_{3/2}$ upper laser level and the higher-energy $^4G_{7/2}$ state is, however, in the same wavelength region, which induces the ESA transition. This phenomenon shifts the gain peak towards the longer wavelength side and severely reduces the efficiency of the $^4F_{3/2}$–$^4I_{13/2}$ stimulated transition [93, 94] when laser operation is attempted at 1.31 μm in silica fiber and even in the phosphate glasses [95].

The ESA spectrum in both wavelength and strength and therefore also in gain is sensitive to glass composition, which allows a positive gain in the 1.3 μm spectral region. Thus the role of glass composition has been examined theoretically and experimentally [96–99]. According to the theoretical analysis [70, 97–99], exponential signal growth, including the ESA process can be characterized by the figure of merit

$$m = C(1 - \sigma_{ESA}/\sigma_e)\sigma_e\tau \qquad (3.39)$$

where C is a constant related to the fiber parameter, σ_e and σ_{ESA} are the stimulated emission and ESA cross section for the signal, and τ is the lifetime of the upper level of the gain transition. The host-dependence properties directly affecting the figure of merit are the ratio of σ_{ESA}/σ_e and $\sigma_e\tau$. This model suggests that fluorozirconate fibers are quite favorable [97–99].

Experiments have also been carried out on the various glass compositions such as fluorozirconate [96, 100, 101] and fluorophosphate glasses [102, 103]. Table 3.6 lists the amplification characteristics of various Nd^{3+}-doped fibers at 1.3 μm spectral region. For instance, Brierley *et al.* [101] obtained the ESA, stimulated emission and the scaled gain spectrum that are shown in Fig. 3.12, using Nd^{3+}-doped fluorozirconate glasses. A gain of 4 ∼ 5 dB was obtained but this was still low at 1.31 μm [101]. In fluorophosphate, the wavelength for maximum gain can only be decreased to ∼1.33 μm.

The second problem to be solved is ASE [101, 105]. Since the stimulated emission cross section and branching ratio for the 1.3 μm transition are approximately four times smaller than those for the 1.06 μm transition, spontaneous emission at 1.06 μm will be amplified more efficiently and saturates the available gain at 1.3 μm. Figure 3.13 shows gain saturation due to ASE, primarily from the strong $^4F_{3/2}$–$^4I_{11/2}$ transition at 1050 nm [101]. A significant improvement can be achieved by filtering out the ASE around 1050 nm [127]

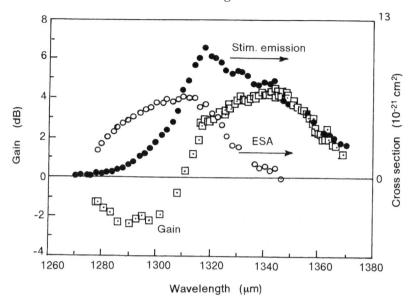

Fig. 3.12.   Excited state absorption spectrum, the $^4F_{3/2}$–$^4I_{13/2}$ emission spectrum and the gain spectrum of Nd^{3+}-doped fluorozirconate fiber. [Reprinted from M. Brierley, S. Carter and P. France, *Electron. Lett.* **26** (1990) 329, copyright (1990) with permission from the Institute of Electrical Engineers.]

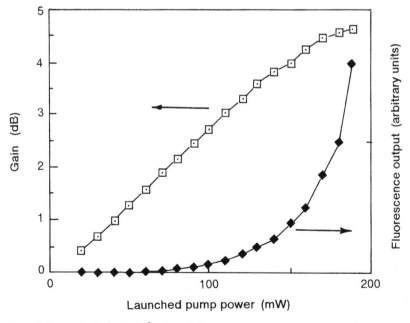

Fig. 3.13.   Gain saturation of Nd^{3+}-doped fluorozirconate fiber at 1.3 μm due to ASE of $^4F_{3/2}$–$^4I_{11/2}$ transition. [Reprinted from M. Brierley, S. Carter and P. France, *Electron. Lett.* **26** (1990) 329, copyright (1990) with permission from the Institute of Electrical Engineers.]

but the best results reported to date are of a net gain of 10 dB at 1.33 μm with a pump power of 150 mW at 800 nm, by using a 10 m long and 1000 ppm Nd^{3+}-doped fluorozirconate fiber [108].

3.3.2.4 1.3 μm amplification in Pr^{3+}-doped fiber

Praseodymium-doped fibers are currently viewed [111–113, 128, 129] as being more promising because their gain is centered at 1.31 μm and there are no competing emission wavelengths. As shown in Fig. 3.11, the four-level $^1G_4$–$^3H_5$ transition has a high branching ratio ($\sim$0.56) and can be used for efficient amplification. The population of $^1G_4$ is achieved by direct pumping of $^1G_4$–$^3H_4$ in the absorption band peaking at 1010 nm. The signal ESA from the upper level of the transition $^1G_4$–$^1D_2$ will modify the gain curve and will reduce the amplifier performance of the $^1G_4$–$^3H_5$ transition. However, the oscillator strength of the $^1G_4$–$^1D_2$ transition is smaller than that of the $^1G_4$–$^3H_5$ transition. In addition, as shown in Fig. 3.14, the peak wavelength is located on the longer wavelength side of the 1.3 μm fluorescence band [130]. Therefore the ESA has only a shorter wavelength tail down to $\sim$ 1290 nm, which moves the peak of the gain curve towards 1.30 μm [111, 112, 128–132].

The major problem with the Pr^{3+}-doped fiber amplifier is the large non-radiative decay rate from $^1G_4$ to the next lower level $^3F_4$, thus reducing the radiative efficiency. As can be seen from the energy diagram in Fig. 3.11, the

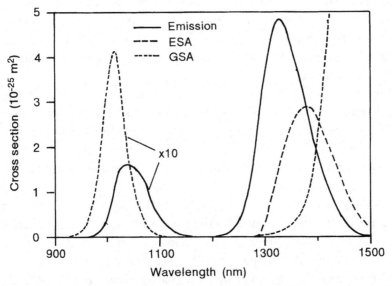

Fig. 3.14. Cross section spectra for the emission, ESA and GSA transitions of Pr^{3+} ion. [Reprinted from B. Pedersen, W. J. Miniscalco and R. S. Quimby, *IEEE Photonics Technol. Lett.* **4** (1992) 446, copyright (1992) with permission from the Institute of Electrical and Electronics Engineers, Inc.]

Table 3.7. *Multiphonon relaxation rates from the $Pr^{3+}: {}^1G_4$ to the ${}^3F_4$ state and quantum efficiency in various oxide and non-oxide glass [Reprinted from D. R. Simons, A. J. Faber, F. Simonis and H. de Waal, 8th International Symposium on Halide Glasses, Perros-Guirec (1992) 448, with permission from Centre National d'Etudes des Telecommunications (CNET)]*

Glass	Composition	Energy gap, ΔE (cm^{-1})	Phonon energy, $\hbar\omega_{ph}$ (cm^{-1})	Relaxation rate, W_{nr} (s^{-1})	Quantum efficiency, η
Oxide	borate	2950[a]	1400	3.9×10^7	7.8×10^{-6}
	phosphate	2950[a]	1200	5.2×10^6	6.0×10^{-5}
	silicate	2950[a]	1100	1.3×10^6	2.3×10^{-4}
	germanate	2878	900	2.7×10^4	1.1×10^{-2}
	tellurite	2950	700	6.9×10^4	4.5×10^{-3}
Fluoride	fluoroberyllate	2950[a]	500	1.3×10^4	2.3×10^{-2}
	fluorozirconate	2941	500	6.4×10^3	4.6×10^{-2}
Sulfide	Al(Ga)–La–S	2950[a]	350	1.0×10^3	2.3×10^{-1}

[a] estimated value.

energy gap between the ${}^1G_4$ and ${}^3F_4$ is approximately 3000 cm^{-1} [111, 112, 132, 133]. Therefore the selection of a host glass with a low phonon energy is very critical in this case. In silica glass, for example, the highest-energy phonon is about 1100 cm^{-1} as shown in Table 3.7 [134] and only three such phonons are needed to bridge the ${}^1G_4$–${}^3F_4$ gap. This leads to a complete quenching of the ${}^1G_4$ level in silica glass and no measurable emission can be expected at 1.3 μm [112, 134].

Emission at 1.31 μm can only be obtained in glasses with low phonon energies. For instance, heavy metal fluoride glasses with low phonon energy (500 cm^{-1}) reduce the non-radiative transition probability substantially relative to fused silica, and make the ${}^1G_4$–${}^3H_5$ transition possible [111–113, 128]. Another nonradiative relaxation process which quenches the ${}^1G_4$ level is believed to be concentration quenching, caused by the cross-relaxation of $({}^1G_4$–${}^3H_5)$–$({}^3H_4$–${}^3F_4)$ and $({}^1G_4$–${}^3H_6)$–$({}^3H_4$–${}^3F_2)$ transfers as shown in Fig. 3.15. The results of fluorescence lifetime measurements indicate that a Pr^{3+} concentration in fluoride fiber of less than 1000 ppm is desirable for the fabrication of efficient Pr^{3+}-doped fluoride fiber amplifiers [106, 132].

Table 3.6 summarizes the amplification characteristics of Pr^{3+}-doped fluoride glasses reported in a number of papers. Figure 3.16 shows a typical example of the gain spectrum of a Pr^{3+}-doped fluoride fiber whose core diameter is 3.3 μm and length is 23 m [106]. The gain was 30.1 dB at

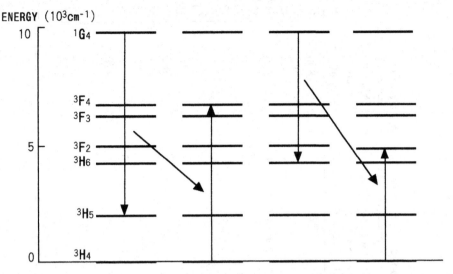

Fig. 3.15.   Possible cross relaxation interaction between Pr^{3+} ions. [Reprinted from Y. Ohishi, T. Kanamori, T. Nishi and S. Takahashi, *IEEE Photonics Technol. Lett.* **3** (1991) 715, copyright (1991) with permission from the Institute of Electrical and Electronics Engineers, Inc.]

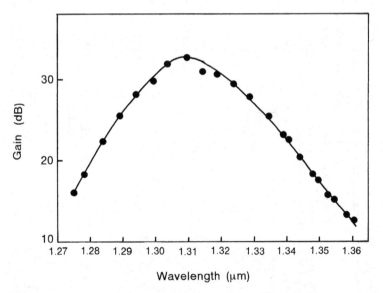

Fig. 3.16.   Typical example of gain spectrum of the Pr^{3+}-doped fluoride fiber amplifier. [Reprinted from Y. Ohishi, T. Kanamori, T. Nishi and S. Takashi, *IEEE Photonics Technol. Lett.* **3** (1991) 715, copyright (1991) with permission from the Institute of Electrical and Electronics Engineers, Inc.]

Table 3.8. (I) Emission characteristics of $^1G_4-^3H_5$ transition of Pr^{3+}-doped glasses with low phonon energy

Base glass type	Fluorescence peak λ_p (μm)	Radiative lifetime τ_R (μs)	Fluorescence lifetime τ_f (μs)	Quantum efficiency η (%)	Branch ratio β	Stimulated emission σ_e (10^{-21} cm²)	Figure of merit $\sigma_e \tau_f$ (10^{-24} cm²s)	Max. photon energy (cm⁻¹)	Ref.
R₂O–ZnO–TeO₂	1.33	475	19	4.0				800	146
PbO–Bi₂O₃–Ga₂O₃	1.32	599	53	9.0	0.46		0.32	500	157
PbO–Bi₂O₃–Ga₂O₃	–	805	202	36				650	159
ZrF₄-fluoride	1.322	3240	110	3.4	0.64	3.48	0.38	580	111
ZrF₄-fluoride		2484	110	4.0	0.6		0.36		142
ZrF₄–PbF₂				3.1 ~ 3.4	0.64–0.65	3.73 ~ 4.06			138
PbF₂–InF₃			172	6.1	0.67	3.89			122
InF₃			158 ~ 180	5.1 ~ 6.2	0.65	3.79 ~ 3.90			138
InF₃–GaF₃			175						110
AlF₃ fluoride	1.32		148					620	141
HfF₄–BaF₂–LaF₃–AlF₃			114					582	153
HfF₄–BaF₂–LaF₃			134					581	153
InF₃–GaF₃		2170	186	8.6	0.64	4.4			123
InF₃–GaF₃				8.1	0.64	4.4			125
CdF₂–NaCl–BaCl		2870	325	11.5			0.77	370	147
CdCl₂		2460	80	3					152
CdCl₂–CdF₂		2460	105	4					152
CdX₂–BaX₂–NaX (X = F, Cl)		216						236	153

Table 3.8. (II) *Emission characteristics of $^1G_4-^3H_5$ transition of Pr^{3+}-doped glasses with low phonon energy*

Base glass type	Fluorescence peak λ_p (μm)	Radiative lifetime τ_R (μs)	Fluorescence lifetime τ_f (μs)	Quantum efficiency η (%)	Branch ratio β	Stimulated emission σ_e (10^{-21} cm^2)	Figure of merit $\sigma_e\tau_f$ (10^{-24} cm^{2}s)	Max. photon energy (cm^{-1})	Ref.
Ge–Ga–S	1.344	511	360	70	0.58	13.3	4.79	250–350	139
CsCl–GaCl$_3$–Ga$_2$S$_3$	1.312	10400	2460	24	0.42	19.8	4.90	325	140
(GeS$_2$)$_{80}$(Ga$_2$S$_3$)$_{20}$		540	320	59	0.5	–	2.5		142
GeS$_x$($x=2\sim3$)		403	358	90				475	145
Ge–La–S	1.34	510	295	58	0.52		2.5	425	143
As–S			250						144
GeS–Ga$_2$S$_3$	1.30		200	60				~345	149
La–Ga–S	1.334	500	300			10.5			151
Ga$_2$S$_3$: La$_2$S$_3$	1.334	700	180	26					152
Ga$_2$S$_3$: La$_2$S$_3$	1.34	520	280	53				300	148,150
Ge–S–I	1.33		368	82				340	154
Ge–Ga–S			377	93					155
Ge–As–S			322	12					155
Ge–S–I			360	82					155
As$_2$S$_3$			250						155
As$_2$S$_3$			277						155
GaS$_{1.5}$LaS$_{1.5}$/CsCl			465						158
Ga–Na–S	1.34		370	56	0.57	10.8	4.0		125
Ge–Ga–S		604	299	50					158

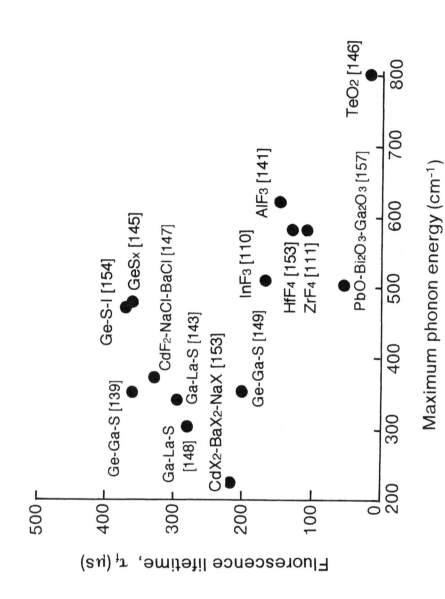

Fig. 3.17. Fluorescence lifetime of $^1G_4$ as a function of maximum phonon energy for various host glasses.

1.309 μm with a pump of 800 mW at 1.107 μm. High gains have also been reported recently at 1.31 μm with a pump of 300 mW at 0.98 μm [118], with a pump of Nd:YLF laser [120], laser diode [109, 114] and master oscillator/ power amplifier laser diode [115, 121].

In efforts to improve the very low absorption cross section of $^1G_4$, co-doping of Yb^{3+}/Pr^{3+} and energy transfer between Yb^{3+} and Pr^{3+} has been examined recently [109, 118, 135]. A Yb-doped fiber laser and a channel waveguide GGG laser operating at around 1.02 μm have also been proposed as the pump source [136, 137].

Although the low phonon energy of fluorozirconate glass certainly reduces the nonradiative transition probability of $^1G_4$, the fluorescence lifetime is approximately 110 μs, which is relatively short compared with the calculated radiative lifetime of 3.2 ms. The quantum efficiency for fluorescence is only 3%. To improve the efficiency of the Pr^{3+}-doped amplifier, other glass hosts with much lower phonon energy have been investigated to further the multi-phonon relaxation rate. Table 3.8 summarizes the radiative and nonradiative characteristics for $^1G_4$ levels of Pr^{3+}-doped glasses with low phonon energy.

Among the fluoride-based glasses, it is known that InF_3-based host glasses have a lower phonon energy than that of ZrF_4-based glasses [138, 152]. The fluorescence lifetime of the $^1G_4$ level in the InF_3-based host was ∼180 μs, which means that the lifetime is 64% longer and thus the quantum efficiency was approximately twice those of ZrF_4-based glasses [138]. A single-pass gain coefficient of 0.36 dB/mW for PbF_2/InF_3-based fluoride glass [122] and both 0.29 dB/mW [123] and 0.65 dB/mW [124] for InF_3/GaF_3-based fluoride fiber ($\tau_f = 175$ μs) has been achieved. AlF_3-based fluoride glasses [160] were also reported to have a long fluorescence lifetime (148 μs) [141].

Further performance improvements can be obtained by using host glasses with much lower photon energies such as mixed halides and chalcogenides [143, 144, 152], as shown in Table 3.8. Figure 3.17 shows the fluorescence lifetime as a function of maximum photon energy for various host glasses. Due to the low phonon energy of these glasses, the multiphonon relaxation rate of the nonradiative transition from $^1G_4$ is low and results in an emission lifetime τ_f of 300 ∼ 380 μs, which is almost three times longer than that observed in oxide and fluoride glasses and leads approximately to a quantum efficiency of ∼60% [143, 151]. Y. Ohishi *et al.* [144] calculated that a gain coefficient of more than 1 dB/mW could be obtained using chalcogenide fibers. By using a strong local electric field around Pr^{3+}, the high refractive index of these glasses ($n > 2$) induces a high radiative transition rate and hence a large cross section for the stimulated emission σ_e. The product of $\sigma_e \tau_f$, which determines the small signal gain coefficient or pump power efficiency of a fiber amplifier

[70, 98], is $2.5 \sim 4.8 \times 10^{-24}$ cm^2 s [139, 142], i.e. one order of magnitude higher than that of ZrF$_4$-based fluoride glasses (0.4×10^{-24} cm^2 s) [106]. A drawback of the chalcogenide glasses is due to the peak emission and hence the expected gain peak being shifted away from the low-loss window to the longer wavelength of 1330 nm [142], as shown in Fig. 3.18.

Most recently, Pr^{3+}-doped Ga$_2$S$_3$-Na$_2$S (66 : 34 mol%) fiber has been fabricated from a preform prepared by an extrusion method and the largest reported gain coefficient of 0.81 dB/mW has been achieved at 1.34 μm [125] and 0.50 dB/mW at 1.31 μm [161], as shown in Fig. 3.19.

3.3.2.5 1.3 μm amplification in Dy^{3+}-doped glasses

The third candidate for 1.3 μm amplification is the Dy^{3+} ion, which is also known to exhibit an emission feature near 1.3 μm. The energy level diagram for the Dy^{3+} ion is shown in Fig. 3.11. The transition of interest for 1.3 μm emission is from the ^{6}H$_{9/2}$: ^{6}F$_{11/2}$ doublet to the ground state ^{6}H$_{15/2}$. Absorption bands of ^{6}H$_{9/2}$, ^{6}H$_{5/2}$, and ^{6}F$_{5/2}$ exist at 1.25, 0.90 and 0.80 μm, respectively, all with a large cross section, which makes efficient pumping of the emitting level possible [162]. However, the energy gap from this doublet to the next lower lying level ^{6}H$_{11/2}$ is relatively small ($1800 \sim 2000$ cm^{-1}) [162, 163, 165]. For example, only two such phonons need to be emitted to

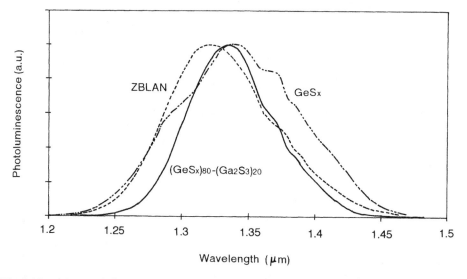

Fig. 3.18.   Measured fluorescence spectra in the second telecommunication window for Pr^{3+}-doped chalcogenide glasses. [Reprinted from D. R. Simons, A. J. Faber and H. de Waal, *Opt. Lett.* **20** (1995) 468, copyright (1995) with permission from the Optical Society of America.]

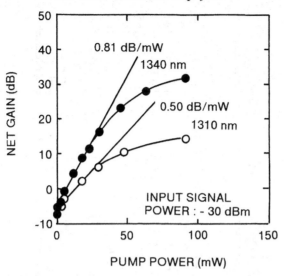

Fig. 3.19. Gain coefficient as a function of pump power for Pr^{3+}-doped chalcogenide fiber. [Reprinted from E. Ishikawa, H. Yanagita, H. Tawarayama, K. Itoh, H. Aoki, K. Yamanaka, K. Okada, Y. Matsuoka and H. Toratani, *Technical Digest, Optical Amplifiers and Their Applications*, Vail (1998) 216, copyright (1998) with permission from the Optical Society of America.]

bridge the energy gap in a fused silica host, in which emission at 1.3 μm is virtually undetectable [163].

Several basic spectroscopic properties including lifetime, radiative transition probabilities and quantum efficiency, were evaluated primarily in chalcogenide with the lowest photon energy. Table 3.9 summarizes the results that appear in the papers. An effective branching ratio of 0.98 is predicted, compared with 0.60 for Pr^{3+}-doped chalcogenide glasses [162, 163, 165]. Another advantage of the Dy^{3+}-doped chalcogenide (Ge : Ga : S) glass is the shorter peak emission wavelength (1320 nm) which is close to the 1310 nm window [165], compared with the Pr^{3+}-doped chalcogenide glasses.

3.3.3 Fiber laser oscillators

A typical configuration for a fiber laser oscillator is schematically shown in Fig. 3.20. Resonant structures for oscillator cavities have been formed by butting cleaved fiber ends to dielectric mirrors [62] or by the application of the dielectric coating directly onto the fiber ends [167]. A typical example of the output power as a function of pump power for the fiber laser is shown in Fig. 3.21 [62]. A great number of continuous wave operations, Q-switching and mode-locking

Table 3.9. *Emission characteristics of $(^6H_{9/2}, {}^6F_{11/2}) - {}^6H_{15/2}$ transition in Dy^{3+}-doped chalcogenide glasses*

Glass type	$\Delta\lambda$ (nm)	λ (μm)	τ_R (μs)	τ_f (μs)	β (%)	η (%)	σ_e (10^{-20} cm^2)	A (s^{-1})	Intensity Parameter (10^{-20} cm^2)			Refs
									Ω_2	Ω_4	Ω_6	
Ge$_{25}$Ga$_5$S$_{70}$	85	1.34	227	38	90.5	16.8	4.35	3990	11.9	3.58	2.17	163
Ge$_{30}$As$_{10}$S$_{60}$		1.334	2220	728	94.0			6132	10.53	3.17	1.17	164
Ge$_{25}$Ga$_5$S$_{70}$		1.336	3290	1130	91.1			3979	11.86	4.00	1.47	164
Ga–La–S	85	1.320		~59		19.0		3300				165
Ga$_2$S$_3$–La$_2$S$_3$	83		203	59	93	29	3.8					162
70Ga$_2$S$_3$30La$_2$S$_3$				25	93	6.9		2594				166
Ge–As–S				20	90	9			11.3	1.11	2.75	155
Ge–S–I				45	91.6	12			7.26	1.52	0.92	155
									U(2)	U(4)	U(6)	
									0.9394	0.8465	0.4078	166

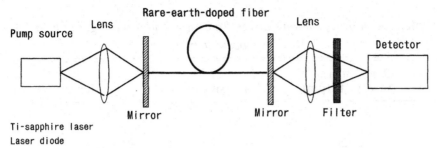

Fig. 3.20. Typical configuration of fiber laser oscillator.

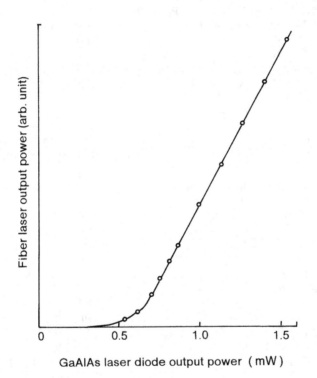

Fig. 3.21. Typical example of the output power as a function of pump power. [Reprinted from R. J. Mears, L. Reekie, S. B. Poole and D. N. Payne, *Electron. Lett.* **21** (1985) 738, copyright (1985) with permission from the Institute of Electrical Engineers.]

operations have already been demonstrated at room temperature in fiber lasers for almost all of rare-earth elements in fluoride-based and silica-based fibers. Table 3.10 summarizes rare-earth ions that have exhibited laser oscillations in glass fiber form and includes sensing ions, the initial and final energy state, and

Table 3.10. *Rare-earth ions, lasing transition, and wavelength of fiber laser oscillator*

Ion	Range of lasing wavelength (µm)	Transition	References
Pr^{3+}	0.490–0.492	$^3P_0 \to {}^3H_4$, $^1I_6 \to {}^3H_5$	170, 171
(Yb/Pr)		$(^*)^3P_0 \to {}^3H_4$, $^1I_6 \to {}^3H_4$	175, 176, 273–276
	0.520	$^3P_1 \to {}^3H_5$, $(^*)^3P_1 \to {}^3H_5$	170, 171, 175, 275, 276
	0.521	$^3P_0 \to {}^3H_5$	179
	0.599–0.618	$^1I_6$, $^3P_0 \to {}^3F_2$, $^3H_6$	169–171, 175, 180
		$(^*)^1I_6$, $^3P_0 \to {}^3F_2$, $^3H_6$	
	0.631–0.641	$^3P_0 \to {}^3F_2$, $(^*)^3P_0 \to {}^3F$	169–171, 175, 178–180, 275
	0.690–0.703	$^3P_0 \to {}^3F_3$, $^3P_1 \to {}^3F_4$	169
	0.707–0.725	$^3P_0 \to {}^3F_4$	169, 171
	0.880–0.886	$^3P_1 \to {}^1G_4$	169
	0.902–0.916	$^3P_0 \to {}^1G_4$	169
	1.048	$^1D_2 \to {}^3F_{3,4}$	174
	1.294–1.312	$^1G_4 \to {}^3H_5$	112, 168, 172, 173, 177, 265
Nd^{3+}	0.381	$(^*)^4D_{3/2} \to {}^4I_{11/2}$	190, 271
	0.412	$(^*)^2P_{3/2} \to {}^4I_{11/2}$	271
	0.90–0.945	$^4F_{3/2} \to {}^4I_{9/2}$	93, 183, 282
	1.049–1.088	$^4F_{3/2} \to {}^4I_{11/2}$	59–61, 93, 167, 181, 182, 184–186, 188, 191, 264, 272, 278
	1.3–1.36	$^4F_{3/2} \to {}^4I_{13/2}$	95, 102, 184–189, 265
Sm^{3+}	0.651	$^4G_{5/2} \to {}^6H_{9/2}$	192
Ho^{3+}	0.550	$(^*)^3F_4$, $^5S_2 \to {}^3I_8$	196, 257, 281
(Tm/Ho)	0.753	$(^*)^3F_4$, $^5S_2 \to {}^3I_7$	196
	1.195	$^5I_6 \to {}^5I_8$	197
	1.38	$^5S_2$, $^5F_4 \to {}^5I$	193
	2.04–2.08	$^5I_7 \to {}^5I_8$	193, 194, 199–201
	2.83–2.95	$^5I_6 \to {}^5I_7$	188, 195
	3.2	$^5F_4$, $^5S_2 \to {}^5F_5$	283
	3.95	$^5I_5 \to {}^5I_6$	198, 280
Er^{3+}	0.54–0.546	$(^*)^4S_{3/2} \to {}^4I_{15/2}$	215, 216, 218
(Yb/Er)	0.981–1.004	$^4I_{11/2} \to {}^4I_{15/2}$	211, 220
	1.534–1.603	$^4I_{13/2} \to {}^4I_{15/2}$	181, 203–207, 209, 210, 212, 213, 220–224, 259–261, 277
	1.66	$^4I_{11/2} \to {}^4I_{9/2}$	213
	1.72	$^4S_{3/2} \to {}^4I_{9/2}$	213, 219
	2.69–2.83	$^4I_{11/2} \to {}^4I_{13/2}$	188, 202, 208, 214, 219, 263
	3.483	$^4F_{9/2} \to {}^4I_{9/2}$	217, 262
Tm^{3+}	0.455	$(^*)^1D_2 \to {}^3H_4$	231, 258
(Yb/Tm)	0.480–0.482	$(^*)^1G_4 \to {}^3H_6$	231, 241, 250, 252, 267, 279
	0.650	$(^*)^1G_4 \to {}^3F_4$	241
	0.803–0.810	$(^*)^3F_4 \to {}^3H_6$, $^3F_4 \to {}^3H_6$	229, 235, 246
	1.445–1.51	$^3F_4 \to {}^3H_4$, $^1D_2 \to {}^1G_4$	229, 231, 236, 239, 243, 248, 266, 268, 269

Table 3.10. (*cont.*)

Ion	Range of lasing wavelength (μm)	Transition	References
	1.82–2.078	$^3H_4 \rightarrow {}^3H_6$	226, 228–230, 232–234, 236, 238, 240, 244, 245, 247, 249, 268
	2.2–2.5	$^3F_4 \rightarrow {}^3H_5$	225, 227, 229, 236–238, 251
Yb^{3+}	0.98–1.14	$^2F_{5/2} \rightarrow {}^2F_{7/2}$	136, 242, 253–256, 270

*: Up-conversion

laser wavelengths reported up to the present. Figure 3.22 summarizes the energy levels, transitions and approximate wavelengths of the trivalent lanthanide ion lasers in fiber form, which appear in Table 3.10. Figure 3.23 also summarizes the spectral range of rare-earth fiber lasers reported up to date.

The laser oscillations summarized in Table 3.10 and Fig. 3.22 include the excitation process of up-conversion, which has attracted a great deal of interest due to the simplicity of this approach which is used to convert efficient and high-power infrared laser diodes to visible light, especially in the blue spectral region. In the fiber laser oscillator, the confinement of pump radiation in a small core fiber leads to a rapid multiple step pumping of high-lying metastable energy levels. Several up-conversion pumped visible-laser systems that are based on lanthanide-doped fluorozirconate glass fiber have been reported (see refs. in Table 3.10). Typical up-conversion processes are also summarized in Fig. 3.24, with the approximate energy levels.

The broad range of potential operating wavelengths provided by fiber laser oscillators makes them useful for a number of applications as light sources in spectroscopic measurement, optical sensing and medical applications.

The operation of a high-grade navigation fiber gyroscope requires an optical source with a broadband spectrum (10 ~ 20 nm) to reduce coherent errors due to back scattering and polarization cross-coupling [284–286]. The super luminescent diode (SLD) [287], which is currently employed, generally has relatively low coupling to single-mode fibers, spectrum sensitivity to optical feedback and more seriously, wavelength sensitivity to temperature. Recently a new broadband source made of a superfluorescent fiber laser and doped with various rare-earth ions such as Nd [285, 286, 288], Yb [289], Pr [290], Er [291, 292] and Sm [293] was proposed and a fiber gyroscope experiment has been demonstrated. A single mode superfluorescent Er/Yb optical fiber source capable of generating more than 1 W of output power was also recently reported [294].

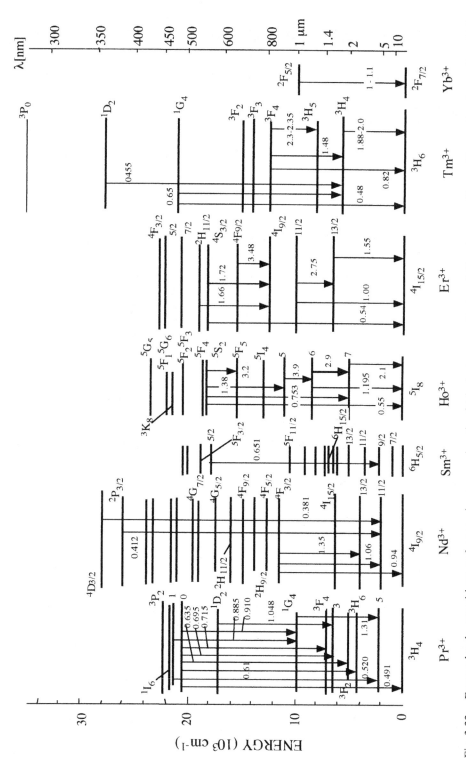

Fig. 3.22. Energy levels, transitions and approximate wavelength of trivalent rare-earth doped fiber laser oscillator. (Energy levels are from [18], transitions (μm) are from Table 3.10.)

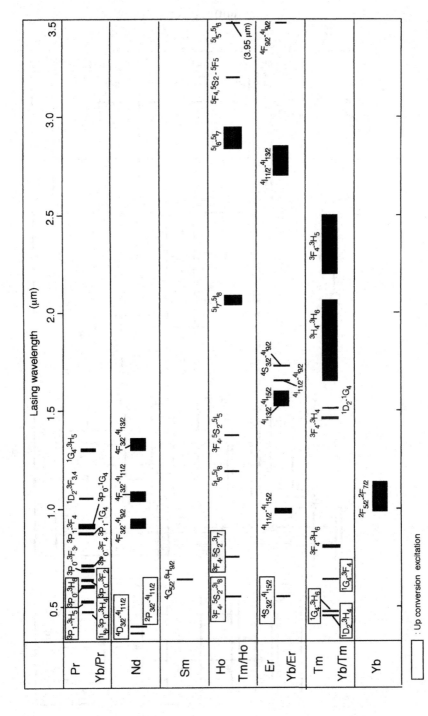

Fig. 3.23.   Spectral range of rare-earth fiber laser.

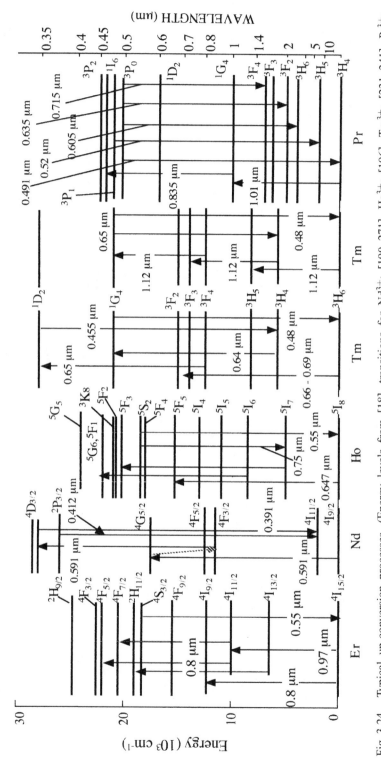

Fig. 3.24. Typical up-conversion processes. (Energy levels from [18], transitions for Nd³⁺: [190, 271], Ho³⁺: [196], Tm³⁺: [190, 271], Tm³⁺: [231, 241], Pr³⁺: [170, 171].)

Fiber lasers doped with Ho^{3+}, Tm^{3+} and Er^{3+} (see Table 3.10 and Figs. 3.22 and 3.23) operate at around 2–3 µm. For instance, CW operation in the 2.75 µm region was established on the $^4I_{11/2} \sim {}^4I_{13/2}$ transition in an Er^{3+}-doped fiber at the pumping wavelengths of 650 and 798 nm ($^4I_{15/2}$–$^4F_{9/2}$, $^4I_{9/2}$) and by an argon ion laser 488 nm into the $^4F_{7/2}$ level or 514.5 nm into the $^2H_{11/2}$ level [188, 202]. It is usually difficult to achieve on the transition $^4I_{11/2}$–$^4I_{13/2}$ since in most host materials the lifetime of the $^4I_{11/2}$ upper lasing level is shorter (approx. 5.2 ms) than that of the $^4I_{13/2}$ lower lasing level (approx. 9.4 ms), making it a self-terminating system. There is concurrent lasing at 1.55 µm from the $^4I_{13/2}$ to the ground state [295], a strong ESA of the respective pump at 476.5 nm from the $^4I_{13/2}$ to the $^4G_{7/2} : {}^2K_{15/2} : {}^4G_{9/2}$ band [202], or at 650 nm and 800 nm into the $^4F_{3/2}$ and $^4F_{5/2}$ or $^2H_{11/2}$ [188] which sufficiently depopulates the $^4I_{13/2}$ level to allow population inversion to be maintained between the $^4I_{11/2}$ level and the $^4I_{13/2}$ level and hence allows CW lasing [202]. This wavelength region is of interest as providing potential sources for molecular-absorption spectroscopy, eye-safe laser radar, monitoring of important molecules such as water, carbon dioxide, hydrogen fluoride, and offers possibilities for sensor applications [236]. Medical applications of long-wavelength lasers take advantage of the water absorption peaks centered at wavelengths of 1.9 and 2.9 µm. Lasers at a wavelength of approximately 2 µm are eye-safe lasers that get absorbed before reaching the retina. Light at this wavelength is also strongly absorbed by water and does not penetrate beyond the outer layers of tissue. They are thus useful for tissue coagulation and allow cutting without bleeding [296]. They can also be used for tissue excision, removal of arterial plaque, cutting bone and drilling teeth [297].

Mid-infrared fiber lasers operating at a wavelength of 4.0 µm also offer numerous applications since this wavelength is the minimum attentuation wavelength of the atmosphere [198].

Erbium-doped fibers seem to offer the best possibility for sources in the useful 1.55 µm telecommunications window [203, 205, 206, 212]. Short-pulse fiber lasers at 1.55 µm make attractive sources for long-range high-spatial resolution-distributed temperature sensors that are based on the technique of optical time domain reflectometry (OTDR) [298].

Yb^{3+}-doped fiber lasers operating on the $^2F_{5/2}$–$^2F_{7/2}$ transition are attractive owing to their wide pumping range extending from 0.8 to 1.0 µm and are efficient means of generating a laser output from 975 to 1180 nm. Moreover, they have only one excited level with a negligible multiphonon relaxation rate in most materials [299]. There are no problems with excited-state absorption of either lasing or pumping photons. Yb^{3+} lasers operating at wavelengths beyond 1 µm are essentially four-level systems, because the energy level diagram shows the three energy levels of the $^2F_{5/2}$ manifold and the four levels of the $^2F_{7/2}$ manifold [253]. Due to the host dependence of Stark splitting, spectroscopy of Yb^{3+} and hence lasing characteristics depend on the host materials. In comparison with silica, the ZrF_4-based fluoride glass is a better host in terms of the emission wavelength but, in contrast, is not so good in terms of the pump wavelength [136]. A Yb^{3+}-doped fiber laser has been used to provide the pump wavelength, 1020 nm, required for pumping a Pr^{3+}-doped ZBLAN fiber [136].

3.3.4 Recent progress

Rare-earth doped chalcogenide glasses represent a group of materials that provide new opportunities for applications in fiber-amplifier and mid-infrared laser devices. Compared with other glass hosts, chalcogenide glasses have some unique properties, as they typically possess a high refractive index and a very low phonon energy. As a result, a large stimulated emission cross section and a low nonradiative relaxation rate can be expected, which enhances the probabilities of observing new fluorescent emission that is normally quenched in oxide glasses. This makes chalcogenide glasses extremely attractive as high-quantum efficiency host materials for rare-earth ions. In addition, because of their good chemical durabilities and good glass-forming abilities, chalcogenide glasses that have been doped with rare-earth ions show promise as fiber amplifier and fiber oscillator materials. Chalcogenide glasses doped with (Nd^{3+}) [159, 278, 300, 301], (Er^{3+}) [148, 302–305], (Pr^{3+}) [159, 306], (Dy^{3+}) [164, 166], and (Tm^{3+}, Ho^{3+}) [307] have recently been studied as candidates for fiber laser devices.

There is currently much interest in some Pr^{3+}-doped transparent glass ceramics which consist of an oxyfluoride for 1.3 μm optical communications. These glass ceramics contain fluoride nanocrystals with a diameter of approximately 15 nm embedded in a primarily oxide glass matrix, which combine the optical advantages of a fluoride host with the mechanical and chemical advantages of oxide glass [308]. The measured quantum efficiency η for Pr^{3+} in the glass ceramics (SiO_2–Al_2O_3–CdF_2–PbF_2–YF_3) is significantly higher than the value of 2.4% obtained for the fluorozirconate host. This high η is due to the Pr^{3+} ions primarily incorporated into the fluoride nanocrystals [309, 310]. Similar glass ceramics that have been doped with (Er^{3+}) [312, 313], (Yb^{3+}) [312], (Tm^{3+}) [314] and (Eu^{3+}) [311] have also been studied for potential application in frequency upconversion.

High-peak-power fiber lasers have also been the subject of much interest during the last few years for various applications in micro-machining. These applications require high intensities of the pump light from multimode diode laser arrays. Such a power scaling in conventional rare-earth-doped monomode fibers is, however, limited because of a poor coupling efficiency between the laser diode array and the single mode core of the fiber. A large fiber diameter which is matched to the emitting area of the high power diode laser arrays increases the efficiency, but leads to multimode operation and reduced beam quality. To overcome these problems, clad-pumping techniques have been developed [315–317].

A clad-pumping fiber consists of a rare-earth-doped single mode inner core (for instance, 5 μm) enclosed by an undoped multimode outer core (or inner clad) of much larger area (for instance, a rectangular shape with dimensions of 150 μm × 75 μm) [318, 319]. The pump light couples efficiently with the multimode outer core and is absorbed whenever light crosses the inner core. Up to 100 W of pump power light can be launched into the outer core. Nd- or Yb-doped clad-pumping fiber lasers have generated output powers in excess of 10 W for CW operation [317–320]. Several kW of peak power and up to 500 mW average power have been obtained by Q-switching operations [318, 319].

3.4 Waveguide lasers and amplifiers

3.4.1 Fabrication of channel waveguide structure

Glass is considered to be a good candidate for passive integrated optical components with low cost and low propagation losses. To date a number of glass waveguide devices such as star couplers and access couplers have been successfully fabricated by a variety of methods. The rare-earth-doped glass waveguide may offer an unusual method of realizing more compact and efficient laser devices that would be difficult to duplicate with rare-earth-doped fibers. In particular, Nd and Er-doped waveguide lasers and amplifiers, fabricated by a variety of methods, have received increasing attention in recent years, because they can be applied not only in optical amplification in the 1.3 and 1.5 μm window but also in potentially stable and high-power laser sources. Early developments in Nd^{3+}-doped and Er^{3+}-doped glass waveguide lasers and amplifiers are reviewed by E. Lallier [321].

Ion exchange is a well-known and widely used fabrication process for waveguide structure [322–326] because this technique permits low propagation losses below 0.1 dB/cm, and the use of commercially available and excellent laser glass substrates. The channel waveguides were formed in the laser glass substrate, for instance, by the usual ion-exchange process [327–330] or by diffusion of an Ag thin film as an ion source [331, 332].

Rare-earth-doped ridge-shape waveguide lasers have also successfully been fabricated [333–336] on Si substrates by flame hydrolysis deposition (FHD) and reactive ion etching (RIE) techniques. Sputtering techniques [337–340], electron beam vapour deposition [341], the pulsed laser deposition technique [342] and the MCVD method [343] instead of FHD have also been used for fabricating a ridge-shape waveguide laser with single transverse operation. Ion implantation has been applied to waveguide fabrication [344] and also to local doping of rare-earth [345] ions into the glass substrate. The sol-gel method has also been used to make an Er^{3+}-doped planar waveguide structure and up-conversion fluorescences were characterized [346].

For some of the rare-earth-doped glasses that have been optimized for laser applications, the ion-exchange technique is not well suited to waveguide fabrication. Such problems are particularly severe with fluoride glasses, which are the best glass hosts for the most interesting rare-earth ions. In order to overcome these problems, new ion-exchange techniques, such as a treatment in the Li or K organic melts, or anionic exchange by substitution of F^- with OH^-, OD^- or Cl^- [347–352] have been applied to multicomponent fluoride glasses. Diffusion of an AgF nonmetallic film into rare-earth-doped fluoroin-date glass substrates [353], vapour phase deposition of the fluoride glass composition [354], and sol-gel synthesis of fluoride glasses from both metal, organic and inorganic precursor systems [355] have also been investigated for the fabrication of fluoride waveguide devices.

Figure 3.25 shows a typical configuration of a waveguide laser oscillator, together with the channel waveguide structure which was fabricated by an ion-exchange technique and a ridge waveguide structure fabricated by FHD and RIE techniques [335]. A laser cavity is formed by two mirrors directly coated or by butting dielectric mirrors on the cleaved ends of the waveguide. The beam from the diode laser is focused into a fiber and guided into the waveguide.

3.4.2 *Lasing characteristics*

3.4.2.1 *Nd^{3+}-doped waveguide laser*

An early experiment on a glass waveguide laser was reported by M. Saruwatari *et al.* [356] using a Nd-doped glass waveguide made by an ion-exchange technique. They formed a 500 μm sized waveguide in Nd^{3+}-doped borosilicate glass using the field-enhanced ion-exchange technique and realized pulse oscillation under dye laser pumping at 590 nm. It was confirmed in this experiment that the threshold of the waveguide laser decreases to one-half that

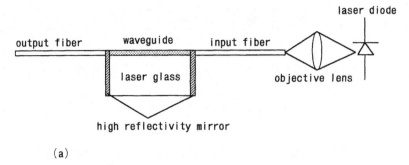

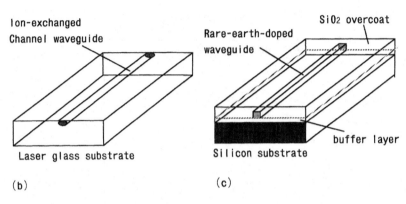

Fig. 3.25. (a) Typical configuration of waveguide laser oscillator and structure of waveguide laser fabricated by (b) ion exchange technique and (c) flame hydrolysis deposition and reaction ion etching technique. [(a) reprinted from H. Aoki, O. Maruyama and Y. Asahara, *IEEE Photonics Technol. Lett.* **2** (1990) 459, copyright (1990) with permission from the Institute of Electrical and Electronics Engineers, Inc. (c) reprinted from T. Kitagawa, K. Hattori, M. Shimizu, Y. Ohmori and M. Kobayashi, *Electron. Lett.* **27** (1991) 334, copyright (1991) with permission from the Institute of Electrical Engineers.]

for the conventional bulk laser. During the last decade there has been intense research on rare-earth-doped glass waveguide lasers as compact and efficient optical devices. For instance, multimode waveguides have been fabricated on a Nd^{3+}-doped phosphate glass substrate with field-enhanced Ag-diffusion technique [331, 332]. The cross section of the waveguide is semicircular with a width of 90 ~ 200 µm. By pumping at 802 nm with a high power LD, CW oscillation was observed at 1054 nm. Maximum output power was higher than 150 mW and slope efficiency was as high as 45% especially in the case of 200 mm core waveguides as shown in Fig. 3.26 [332]. The performance is comparable with that of miniature lasers and may find application as a compact microlaser source.

A nearly single-transverse-mode channel waveguide laser was also fabricated using a field-assisted K^+–Na^+ ion exchange on a Nd^{3+}-doped soda-lime

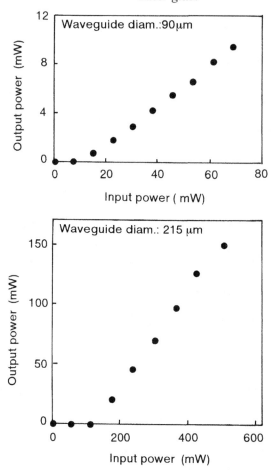

Fig. 3.26. CW lasing characteristics of Nd^{3+}-doped glass waveguide laser at 1054 nm. [Reprinted from H. Aoki, O. Maruyama and Y. Asahara, *IEEE Photonics Technol. Lett*, **2** (1990) 459, copyright (1990) with permission from the Institute of Electrical and Electronics Engineers, Inc.]

silicate glass substrate [327], using FHD and RIE [334, 357], and also using a magnetron-sputtering technique [339]. Almost single mode CW oscillations have been observed in each case.

A narrow linewidth waveguide laser has been demonstrated using an intracore Bragg grating, as distributed Bragg reflectors (DBR). This type of waveguide laser was proposed by S. I. Najafi *et al.* [330] and realized for the first time by J. E. Roman *et al.* [358, 359]. A 5.5 mm long and 0.35 μm period square-wave grating as shown in Fig. 3.27 was produced holographically in photoresist and then etched into the waveguide using argon ion milling. A grating period followed from the Bragg condition

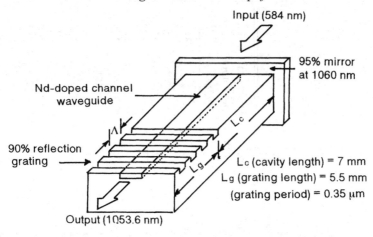

Input (584 nm)

95% mirror
at 1060 nm

Nd-doped channel
waveguide

90% reflection
grating

L_c (cavity length) = 7 mm
L_g (grating length) = 5.5 mm
(grating period) = 0.35 μm

Output (1053.6 nm)

Fig. 3.27. Glass waveguide laser with distributed Bragg reflector. [Reprinted from J. E. Roman and K. A. Winick, *Appl. Phys. Lett.* **61** (1992) 2744, copyright (1992) with permission from the American Institute of Physics.]

$$\varLambda = \lambda_0/2N_{\text{eff}} \qquad (3.40)$$

where λ_0 is the wavelength at peak reflectivity and N_{eff} is the effective index of the guided mode. The laser operated in a single longitudinal mode with a line width less than 37.5 MHz spectrometer resolution.

The 1.3 μm spectral region is an important channel in fiber optic telecom-munications because of the low-loss and favourable dispersion properties of fused-silica fiber. Glass waveguide lasers operating in this wavelength region have also been demonstrated [360, 361]. Table 3.11 summarizes the oscillation characteristics of the Nd^{3+}-doped glass waveguide lasers reported up to date.

3.4.2.2 *Er³⁺-doped waveguide laser*

The Er^{3+}-doped silica waveguide laser operating around 1.6 μm was realized using a low scattering-loss planar waveguide with an Er-doped silica core which was fabricated by the FHD and RIE techniques on a Si substrate [335]. CW lasing was achieved successfully with an incident threshold power of 49 mW for pumping at 0.98 μm. Laser oscillation at around 1.55 μm was also achieved for the ion-exchanged waveguide in Er^{3+}-doped glass [368] and Er^{3+}/Yb^{3+} co-doped glass [372]. More recently, the single-longitudinal mode operation was demonstrated [369] by using an $Er^{3+}-P_2O_5$ co-doped planar waveguide laser with Bragg grating reflectors. Arrays of distributed Bragg-reflector waveguide lasers at 1536 nm in Yb/Er co-doped phosphate glass has also been reported [378].

Table 3.11. *Oscillation characteristics of rare-earth doped glass waveguide lasers*

R.E. ion	Substrate or materials	Waveguide diameter (μm)	Length (mm)	Pumping wavelength (nm)	Lasing wavelength (μm)	Lasing threshold (mW)	Slope efficiency (%)	Maximum output/input (mW)	Ref.
Nd	Borosilicate (IE)	40 × 30	4	595 (Dye)	1.06	23 μJ			356
	Phosphate (IE)	215	5	802 (LD)	1.054	170	45	150/500	332
	Phosphate (IE)	90	5	802 (LD)	1.325	21	0.77		360
	Silica (FHD)	6 × 20	50	800 (Ti)	1.052	150	0.12		334
	Sodalime (IE)		10	528 (Ar)	1.057	31	0.5		327
	BK7 (IE)	~5	17	807 (Ti)	1.058	7.5			328
	Phosphate (SPT)	2 × 10	4	590 (Dye)	1.064	100	6	3/	339
	Sodalime (IE)		12	807 (Ti)	1.057	20	4 ~ 11	0.1/500	362
	Silica (FHD)	6 × 8	40	805 (LD)	1.053	25	1.2		357
	Silica (FHD)	7 × 26	35	807 (Ti)	1.060	143	2.1		336
	Silica (FHD)	12 × 8	59	804 (Ti)	1.053	20	2.6	> 1.0/	363
	Phosphate (IE)			794 (Ti)	1.054	75	3.25	5.2/229	365
	Phosphate (IE)			804 (Ti)	0.906	170			361
	Phosphate (IE)			804 (Ti)	1.057	12	56	210/	361
					1.358	52			361
	Silica (FHD)	25 × 7	45	802 (Ti)	1.053	22	2.5		366

Ion	Host (method)	Dimensions	Length	Pump (source)	n				Ref.
	Silicate (IE)		70	806 (Ti)	1.06	135	1		367
	Sodalime (IE, DBR)		24	584 (Dye)	1.054	50	5.1		359
	Sodalime (IE)		40	815 (Ti)	1.057	26	2.0		374
Er	Silica (FHD)	6 × 8	92	805 (Ti)	1.053	26	1.1		375
	Silica (FHD)	7 × 8	45	980 (Ti)	1.536	250	0.8		364
	Silica (FHD)	6 × 25	64.5	980 (Ti)	1.60	49		1.2/200	335
	GeO$_2$-SiO$_2$ (EB)	10 × 8	36	980 (Ti)	1.53	100 ∼ 150			341
	BK7 (IE)			980 (Ti)	1.544	150	0.55	0.4/270	368
	Silica (FHD, DBR)	8 × 7		976 (Ti)	1.546	60		0.34/300	369
	Silica (FHD, ML)		47	1480, 980 (LD)	1.535				370
Eb/Yb	Borosilicate (IE)		20	977 (LD)	1.544	5	2	1/	371
	Silicate (IE)		8.6	967 (Ti)	1.537	14.8	5.54		372
	Phosphate (IE)	7.3 × 6.4	43	980 (LD)	1.550	65	6		373
	- (IE)		43	980 (LD)	1.536	45	10.6	2.2/70	376
	- (IE)		43	980 (LD)	1.536	65	1		377
Tm	Germanate (II.)		105	790 (Ti)	1.92	167			344

IE, Ion exchange; FHD, Flame hydrolysis deposition; SPT, Sputtering; II, Ion implantation; EB, Electron beam vapour deposition; LD, laser diode; Dye, Dye laser; Ti, Ti sapphire laser; Ar, Ar ion laser; ML, Mode-locking; DBR, Distributed Bragg Reflection.

3.4.2.3 Other rare-earth-doped waveguide lasers

The planar waveguide laser in the 2 μm region was created by implantation of the polished surface of the Tm^{3+}-doped germanate glass with 2.9 MeV He^+ ions [344]. Laser operation of the $^3H_4-^3H_6$ transition of Tm^{3+} ions was observed in this waveguide. The lasing characteristics of Er^{3+} and other rare-earth-doped glass waveguides were listed in Table 3.11.

3.4.3 Amplification characteristics

3.4.3.1 Nd³⁺-doped waveguide amplifier

The first account of the amplification characteristics of an active glass waveguide was reported by H. Yajima *et al.* [337]. They sputtered Nd-doped glass on Corning 7059 glass substrate and pumped them with a Xe flass tube. The increases in power amplitude of a 3 cm-long Nd-doped glass thin waveguide was only 28% (0.36 dB/cm). More recently, thin film waveguides of Nd^{3+}-doped phosphate laser glass [339] and ion-exchanged waveguides in Nd^{3+}-doped laser glass [379] have been investigated extensively. Recent results [380, 381] on ion-exchanged waveguides have reported a 15 dB amplification at 1.06 μm. The amplification characteristics of Nd^{3+}-doped glass waveguides are summarized in Table 3.12.

3.4.3.2 Er³⁺-doped waveguide amplifier

The amplification characteristics of Er^{3+}-doped glass waveguides that were fabricated by using the RF sputter technique were investigated to realize waveguide amplifiers operating at around 1.55 μm for optical communications [340]. By pumping the waveguide with 120 mW of power at 975 nm, the 1.53 μm signal intensity was enhanced by 21 dB. The measured gain was, however, 0 dB for a 2.4 cm long waveguide due to the waveguide loss. Er^{3+}-doped silica-based waveguides fabricated by FHD and RIE on Si substrates have also been developed [382, 392]. The small signal amplification power characteristics of the 19.4 cm-long waveguide are shown in Fig. 3.28. A net gain of 13.7 dB was obtained at a wavelength of 1.535 μm with a pump power of 640 mW. Quite recently, a 15 dB net gain has been achieved on a 4.5 cm long amplifier which was fabricated from Er^{3+}-doped soda-lime glass films by RF sputtering [385, 388].

The optimization of an Er^{3+}-doped silica glass waveguide amplifier was discussed by taking concentration quenching and cooperative upconversion effects into account [393–396]. As a result of these discussions, a gain of 20 dB was calculated with an Er^{3+} ion concentration of $0.4 \sim 0.7$ wt% when a

Table 3.12. *Recent amplification characteristics of rare-earth-doped glass waveguides*

R. E. Ion	Substrate or Materials	Waveguide size (μm)	Pumping wavelength (nm)	Signal wavelength (nm)	Pumping power (mW)	Gain/length (dB/cm)	Ref.
Nd	Barium crown (SPT)	2.7 (slab)	Xe flash	1.06	6.6 J/cm	1.08/3	337
	Barium crown (SPT)	1.2 (slab)	585 (Dye)	1.064	0.1		338
	Phosphate (SPT)	2 × 10	590 (Dye)	1.064	~500	~3/0.4	339
	Soda-lime (IE)	17 × 33	815 (Ti)	1.057	85	3/2.4	374
	Phosphate (IE)	5 × 7	514.5 (Ar)	1.054	140	3.4	379
	Silicate (IE)		532	1.064	100	15/	380, 381
Er	Silica (FHD)	8 × 7	980 (Ti)	1.535	640	13.7/19.4	382
	Silicate (SPT)	1.3 × 6	975 (Ti)	1.53	120	21/2.4	340
	Phosphate		973 (Ti)	1.523		10/0.47	383
	Silicate (IE)	4	528 (Ar)	1.54		2/3.8	384
	Soda-lime (SPT)	1.5 × 3 ~ 10	980	1.53	280	15/4.5	385
	Silica (FHD)	9 × 7	980 (LD)	1.533	99	9.4/23	386
			980 (Ti)		210	16/23	386
	Soda-lime (SPT)	1.4 × 3 ~ 9	980 (Ti)	1.536	80	4.5/6	387
	Soda-lime (SPT)	1.5 × 5 (slab)	980	1.537	280	15/4.5	388
	Phosphate (SPT)	1 × 4	980 (Ti)	1.535	21	4.1/1.0	389
Er/Yb	Borosilicate (IE)		978 (Ti)	1.537	130	9/3.9	390
	Phosphate (IE)		1000	1.530	50	7.3/0.6	391

SPT, Sputtering; IE, Ion exchange; FHD, Flame hydrolysis deposition; Dye, Dye laser; Ti, Ti sapphire laser; Ar, Ar ion laser; LD, Laser diode.

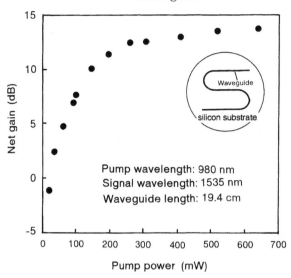

Fig. 3.28.   Amplification characteristics of the 19.4 cm long Er-doped silica waveguide. [Reprinted from T. Kitagawa, K. Hattori, K. Shuto, M. Yasu, M. Kobayashi and M. Horiguchi, *Electron. Lett.* **28** (1992) 1818, copyright (1992) with permission from The Institute of Electrical Engineers.]

20 wt% P_2O_5-co-doped waveguide was used [395]. A 12 dB gain at 150 mW pump power has also been estimated with a 4 cm-long waveguide in high Er^{3+}-doped phosphate glasses [396].

Efficient Yb^{3+} to Er^{3+} energy transfer is a useful mechanism to reduce performance degradation due to the Er^{3+} ion–ion interactions. Numerical calculations based on realistic waveguide parameters demonstrated the possibility of achieving high gain with a short device length. A gain coefficient of $1.5 \sim 3.0$ dB/cm was estimated numerically on the higher concentration Er^{3+}/Yb^{3+} co-doped waveguide amplifiers [397, 398]. A diode-pumped planar Er/Yb waveguide laser has also been demonstrated recently [371]. The amplification characteristics of Er^{3+} and Er^{3+}/Yb^{3+}-doped glass waveguides are listed in Table 3.12.

3.4.4 Devices based on glass waveguide lasers

The growing need for a cost-effective extension of optical networks from the kerb to business premises, or even the home, has made the development of lossless devices highly desirable. One such device is a lossless splitter as shown in Fig. 3.29 (a) and (b), which takes an incoming signal from an optical fibre and divides it between multiple output fibers but maintains the signal strength

(a)

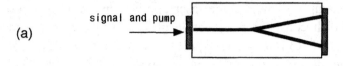

(b)

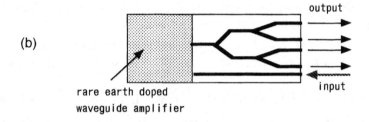

(c)

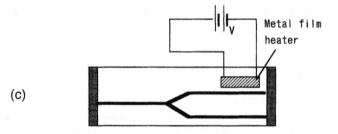

Fig. 3.29.   Devices based on the glass waveguide lasers (a) Y-branch loss compensator, (b) loss-less beam splitter, (c) modulator.

with an integrated optical amplifier. Such a device was built by N. A. Sanford *et al.* [374]. They demonstrated that the 3 dB attenuation suffered by a single output channel of the device owing to uniform splitting is recovered by the absorption of approximately 85 mW of pump power. Lossless splitting over the wavelength range 1534 $\sim$ 1548 nm was also achieved using a Er^{3+}/Yb^{3+} co-doped borosilicate glass waveguide fabricated by Tl^+-ion exchange [390].

A successful system demonstration of a four-wavelength integrated-optics amplifying combiner has been reported [399]. This device is based on an Er^{3+}/Yb^{3+}-doped waveguide amplifier and integrated splitters. The arrangement consists of an all-connectorized 4 $\times$ 1 glass splitter followed by a 4.5 cm-long Er^{3+}/Yb^{3+}-co-doped waveguide amplifier. In the amplifying section, 11.6 dB of net gain for a single-pass configuration and 23 dB for a double-pass was recorded. These results show their potential use in the manufacture of lossless telecommunication devices [399].

Another example of waveguide laser applications is a thermo-optic phase modulator which consists of Y-junction glass waveguide as shown schematically in Fig. 3.29 (c) [400]. The Y-junction waveguide was fabricated by K^+-ion exchange in BK7 glass that was doped with 1.5 wt% Nd oxide. An aluminium electrode with dimensions of 5 mm $\times$ 0.2 mm $\times$ 10 mm and a resistance of approximately 200 Ω was positioned next to one branch of the waveguide. Figure 3.30 shows the response of the laser output waveforms to the applied square-wave voltage of 100 Hz, which was stepped between 0 and 3.3 V, while maintaining a constant pumping level [400]. The Q-switched pulses were long because of the slow response of the thermo-optic modulator. More recently, a tunable Y-branched Er-doped glass waveguide laser was also built [401].

An active [362] and passive [365] Q-switching operation have also been demonstrated using a Nd-doped channel waveguide, which generated Q-switched peak power of 1.2 W and 3.04 W, respectively. A high-power Er^{3+}/Yb^{3+} co-doped glass-waveguide, Q-switched laser which generated [373] 67 W peak power and a 200 ns pulse at a 1 kHz repetition rate has also been described.

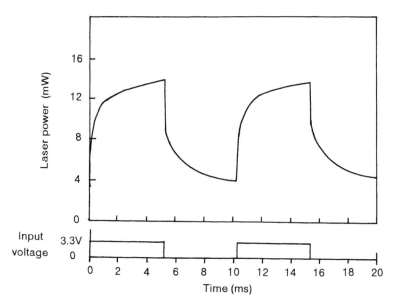

Fig. 3.30.   Lasing power characteristics of the glass waveguide modulator. [Reprinted from E. K. Mwarania, D. M. Murphy, M. Hempstead, L. Reekie and J. S. Wilkinson, *IEEE Photonics Technol. Lett.* **4** (1992) 235, copyright (1992) with permission from the Institute of Electrical and Electronics Engineers, Inc.]

References

1 M. J. Weber: *Handbook on the Physics and Chemistry of Rare Earths Vol. 4, Non-Metallic Compound II*, ed. K. A. Gschneidner, Jr. and L. R. Eyring, Chap. 35, (North-Holland Physics Publishing, Amsterdam, 1979) p. 275.

2 A. Yariv: *Optical Electronics*, Chap. 5 & 6, (Oxford University Press, Oxford, 1991) p. 148, p. 174.

3 W. Koechner: Chap. 1 in *Solid State Laser Engineering*, (Fourth extensively revised and updated edition) (Springer-Verlag, New York, 1996) p. 1, 148, 174.

4 L. A. Riseberg and M. J. Weber: *Progress in Optics*, Vol XIV, ed. E. Wolf, (North Holland, Amsterdam, 1976) p. 89.

5 H. A. Pohl: *Quantum Mechanics for Science and Engineering*, (Prentice-Hall, Inc., Englewood Cliffs, 1967).

6 P. F. Moulton: *Laser Handbook*, Vol. 5, ed. M. Bass and M. L. Stitch, (North Holland Publishers Co., Amsterdam, 1985) p. 203.

7 D. Sutton: *Electronic Spectra of Transition Metal Complexes*, (McGraw-Hill Publishing Company Limited, New York, 1969).

8 D. W. Hall and M. J. Weber: *CRC Handbook of Laser Science and Technology Supplement 1: Laser*, ed. M. J. Weber, (CRC Press, Boca Raton, 1991) p. 137.

9 W. Koechner: Chap. 2 in *Solid State Laser Engineering*, (Springer-Verlag, New York, 1976) p. 32.

10 M. J. Weber: *J. Non-Cryst Solids* **123** (1990) 208.

11 J. A. Caird and S. A. Payne: *CRC Handbook of Laser Science and Technology, Supplement 1: Lasers*, ed. M. J. Weber, (CRC Press, Boca Raton, 1991) p. 3.

12 A. A. Kaminskii: *Laser Crystals*, translation ed. H. F. Ivey, (Springer-Verlag, Berlin, 1981).

13 R. Reisfeld and C. K. Jørgensen: Excited State Phenomena in Vitreous Materials in *Handbook on the Physics and Chemistry of Rare Earth*, Vol. 9, ed. K. A. Gschneidner Jr. and L. Eyring, (North Holland, Amsterdam, 1987) p. 1.

14 R. D. Peacock: *Structure and Bonding* Vol. 22 (Springer-Verlag, Berlin, 1975) p. 83.

15 R. Reisfeld: ibid. p. 123.

16 R. Reisfeld and C. K. Jørgensen: *Lasers and Excited States of Rare Earths*, (Springer-Verlag, 1977).

17 K. Patek: *Glass Lasers*, (Iliffe Books, London, 1970).

18 G. H. Dieke and H. M. Crosswhite: *Appl. Opt.* **2** (1963) 675.

19 W. Krupke: *IEEE Quantum Electron.* **QE-10** (1974) 450.

20 M. J. Weber and R. Jacobs: *Laser Program Annual Report-1974*, (Lawrence Livermore Laboratory, 1975) p. 263.

21 A. J. Dekker: Chap. 6 in *Solid State Physics* (Prentice-Hall, Inc., Englewood Cliffs and Marugen Company Ltd, Tokyo, 1957) p. 133.

22 G. Burns: Vol. 4, Chap. 3 in *Solid State Physics* (Academic Press Inc., New York, 1985), Translation in Japanese, S. Kojima, S. Sawada and T. Nakamura (Tokai Daigaku Shuppankai, 1991) p. 89.

23 B. R. Judd: *Phys. Rev.* **127** (1962) 750.

24 G. S. Ofelt: *J. Chem. Phys.* **37** (1962) 511.

25 R. R. Jacobs and M. J. Weber: *IEEE Quantum Electron.* **QE-12** (1976) 102.

26 W. T. Carnell, P. R. Fields and K. Rajnak: *J. Chem. Phys.* **49** (1968) 4424.

27 W. B. Fowler and D. L. Dexter: *Phys. Rev.* **128** (1962) 2154.

28 L. A. Riseberg and H. W. Moos: *Phys. Rev.* **174** (1968) 429.

29 T. Miyakawa and D. L. Dexter: *Phys. Rev.* **B1** (1970) 2961.

30 M. J. Weber and A. J. DeGroot: *Laser Program Annual Report – 1976* (Lawrence Livermore Laboratory, 1977) p. 2-271.

31 S. E. Stokowski: *Laser program Annual Report – 83* (Lawrence Livermore National Laboratory, 1984) p. 6-40.

32 W. J. Miniscalco: *J. Lightwave Technol.* **9** (1991) 234.

33 M. J. F. Digonnet, M. K. Davis and R. H. Pantell: *Optical Fiber Technol.* **1** (1994) 48.

34 E. Snitzer: *Phys. Rev. Lett.* **7** (1961) 444.

35 D. C. Brown: *High-Peak-Power Nd:Glass Laser Systems*, (Springer-Verlag, Berlin, 1981).

36 M. J. Weber and J. E. Lynch: *1982 Laser Program Annual Report*, (Lawrence Livermore National Laboratory, 1983) p. 7-45, and refs. cited therein.

37 M. A. Henesian, W. T. White, M. J. Weber and S. A. Brawer: *1981 Laser Program Annual Report*, (Lawrence Livermore National Laboratory, 1982) p. 7-33.

38 M. J. Weber, R. A. Saroyan and R. C. Ropp: *J. Non-Cryst. Solids* **44** (1981) 137.

39 T. Izumitani, H. Toratani and H. Kuroda: *Non-Cryst. Solids* **47** (1982) 87.

40 S. Tanabe, T. Ohyagi, N. Soga and T. Hanada: *Phys. Rev.* **B46** (1992–II) 3305.

41 S. Tanabe, T. Hanada, T. Ohyagi and N. Soga: *Phys. Rev.* **B48** (1993–II) 10591.

42 H. Takebe, Y. Nageno and K. Morinaga: *J. Am. Ceram. Soc.* **78** (1995) 1161.

43 C. B. Layne, W. H. Lowdermilk and M. J. Weber: *Phys. Rev.* **B16** (1977) 10.

44 C. B. Layne and M. J. Weber: *Laser Program Annual Report – 1976*, (Lawrence Livermore Laboratory, 1977) p. 2-272.

45 C. B. Layne and M. J. Weber: *Phys. Rev.* **B16** (1977) 3259.

46 J. Lucas and J. L. Adam: *Glastech. Ber.* **62** (1989) 422, and refs. cited therein.

47 M. Shojiya, M. Takahashi, R. Kanno, Y. Kawamoro and K. Kadono: *Appl. Phys. Lett.* **72** (1998) 882.

48 S. E. Stokowski, *CRC Handbook of Laser Science and Technology, Vol. I, Lasers and Masers*, Ed. M. J. Weber, (CRC Press Inc., Boca Raton FL, 1985) p. 215.

49 H. Kuroda, H. Masuko, S. Maekawa and T. Izumitani: *J. Appl. Phys.* **51** (1980) 1351.

50 *Hoya Laser Glass Catalogue E9107 – 13005 D*.

51 L. Prod'homme: *Phys. Chem. Glasses* **1** (1960) 119.

52 F. Reitmayer and H. Schroeder: *Appl. Opt.* **14** (1975) 716.

53 T. Izumitani and H. Toratani: *J. Non-Cryst. Solids* **40** (1980) 611.

54 W. Koechner: *Solid-State Laser Engineering*. Fourth extensively revised and updated edn. Chap. 7, (Springer-Verlag, Berlin, 1996) p. 393.

55 S. E. Stokowski: *Laser Program Annual Report 83*, (Lawrence Livermore National Laboratory, 1984) p. 6-44.

56 S. E. Stokowski: *1986 Laser Program Annual Report*, (Lawrence Livermore National Laboratory, 1987) p. 7-150.

57 M. J. Weber, C. B. Layne and R. A. Saroyan: Laser Program Annual Report – 1975 (Lawrence Livermore Laboratory, 1976) p. 192.

58 C. J. Koester and E. Snitzer: *Appl. Opt.* **3** (1964) 1182.

59 J. Stone and C. A. Burrus: *Appl. Phys. Lett.* **23** (1973) 388.

60 J. Stone and C. A. Burrus: *Appl. Opt.* **13** (1974) 1256.

61 S. B. Poore, D. N. Payne and M. E. Fermann: *Electron. Lett.* **21** (1985) 737.

62 R. J. Mears, L. Reekie, S. B. Poole and D. N. Payne: *Electron Lett.* **21** (1985) 738.

63 S. B. Poole, D. N. Payne, R. J. Mears, M. E. Fermann and R. I. Laming: *J. Lightwave Technol.* **LT-4** (1986) 870.

64 P. Urquhart: *IEE Proceedings* **135** (1988) 385.
65 R. J. Mears and S. R. Baker: *Optical and Quantum Electron.* **24** (1992) 517, and refs. cited therein.
66 M. J. F. Digonnet and B. J. Thompson (ed.): *Selected Papers on Rare-Earth-Doped Fiber Laser Sources and Amplifiers*, SPIE Milestone Series Vol.MS37, (SPIE Optical Engineering Press, 1992).
67 Y. Miyajima, T. Komukai, T. Sugawa and T. Yamamoto: *Optical. Fiber Technol.* **1** (1994) 35, and refs. cited therein.
68 N. S. Kapany: *Fiber Optics, Principles and Applications*, (Academic Press, Inc. New York, 1967).
69 K. J. Beales and C. R. Day: *Phys. Chem. Glass.* **21** (1980) 5.
70 M. J. F. Digonnet and C. J. Gaeta: *Appl. Opt.* **24** (1985) 333.
71 W. A. Gambling: *Glass Technol.* **27** (1986) 179.
72 T. Izawa and S. Sudo: *Optical Fibers: Materials and Fabrication* (KTK Scientific Publications, Tokyo, D. Reidel Publishing Company, Dordrecht, 1987).
73 H. Poignant, C. Falcou and J. le Mellot: *Glass Technol.* **28** (1987) 38.
74 T. Miyashita and T. Manabe: *IEEE J. Quantum Electron* **QE-18** (1982) 1432.
75 J. Nishii, S. Morimoto, I. Inagawa, R. Iizuka, T. Yamashita and T. Yamagishi: *J. Non-Cryst. Solids* **140** (1992) 199.
76 J. E. Townsend, S. B. Poole and D. N. Payne: *Electron. Lett.* **23** (1987) 329.
77 B. J. Ainslie, S. P. Craig, S. T. Davey and B. Wakefield: *Mat. Lett.* **6** (1988) 139.
78 S. Mitachi, T. Miyashita and T. Kanamori: *Electron. Lett.* **17** (1981) 591.
79 S. Mitachi, T. Miyashita and T. Manabe: *Phys. Chem. Glass.* **23** (1982) 196.
80 Y. Ohishi, S. Mitachi and S. Takahashi: *J. Lightwave Technol.* **LT-2** (1984) 593.
81 Y. Ohishi, S. Sakaguchi and S. Takahashi: *Electron. Lett.* **22** (1986) 1034.
82 D. C. Tran, C. F. Fisher and G. H. Sigel: *Electron Lett.* **18** (1982) 657.
83 K. Itoh, K. Miura, I. Masuda, M. Iwakura and T. Yamashita: *J. Non-Cryst. Solids* **167** (1994) 112.
84 R. J. Mears, L. Reekie, I. M. Jauncey and D. N. Payne: *Electron. Lett.* **23** (1987) 1026.
85 K. Arai, H. Namikawa, K. Kumada and T. Honda: *J. Appl. Phys.* **59** (1986) 3430.
86 E. Desurvire, J. R. Simpson and P. C. Becker: *Opt. Lett.* **12** (1987) 888.
87 T. J. Whitley: *Electron. Lett.* **24** (1988) 1537.
88 R. I. Laming, S. B. Poole and E. J. Tarbox: *Opt. Lett.* **13** (1988) 1084.
89 R. I. Laming, M. C. Farries, P. R. Morkel, L. Reekie, D. N. Payne, P. L. Scrivener et al.: *Electron. Lett.* **25** (1989) 12.
90 E. Snitzer, H. Po, F. Hakimi, R. Tumminelli and R. C. McCollum, *Tech. Dig. Opt. Fiber Commun. Conf.* (Optical Society of America, Washington, DC, 1988) PD2.
91 M. Nakazawa, Y. Kimura and K. Suzuki: *Appl. Phys. Lett.* **54** (1989) 295.
92 K. Inoue, H. Toba, N. Shibata, K. Iwatsuki, A. Takada and M. Shimizu: *Electron Lett.* **25** (1989) 594.
93 I. P. Alcock, A. I. Ferguson, D. C. Hanna and A. C. Tropper: *Opt. Lett.* **11** (1986) 709.
94 P. R. Morkel, M. C. Farries and S. B. Poole: *Opt. Commun.* **67** (1988) 349.
95 S. G. Grubb, W. L. Barnes, E. R. Taylor and D. N. Payne: *Electron. Lett.* **26** (1990) 121.
96 L. J. Andrews: Technical Digest, Topical Meeting on Tunable Solid State Lasers (Optical Society of America, Washington, DC, 1987) WB1.
97 W. J. Miniscalco, B. Pedersen, S. Zemon, B. A. Thompson, G. Lambert, B. T. Hall,

et al.: *Mat. Res. Soc. Symp., Proc.* Vol. 244 (Materials Research Society, 1992) p. 221.

98 M. L. Dakss and W. J. Miniscalco: *IEEE Photonics Technol. Lett.* **2** (1990) 650.

99 M. L. Dakss and W. J. Miniscalco: *SPIE* vol. **1373** Fiber Laser Sources and Amplifiers II (1990) p. 111.

100 J. E. Pedersen and M. C. Brierley: *Electron. Lett.* **26** (1990) 819.

101 M. Brierley, S. Carter and P. France: *Electron. Lett.* **26** (1990) 329.

102 E. Ishikawa, H. Aoki, T. Yamashita and Y. Asahara: *Electron. Lett.* **28** (1992) 1497.

103 S. Zemon, B. Pedersen, G. Lambert, W. J. Miniscalco, B. T. Hall. R. C. Folweiler *et al.*: *IEEE Photonics Technol. Lett.* **4** (1992) 244.

104 T. Nishi, M. Shimizu and S. Takashi: *The 2nd Meeting on Glasses for Optoelec-tronics*, January 18 (1991) 02.

105 Y. Miyajima, T. Sugawa and T. Komukai: *Electron. Lett.* **26** (1990) 1397.

106 Y. Ohishi, T. Kanamori, T. Nishi and S. Takahashi: *IEEE Photonics Technol. Lett.* **3** (1991) 715.

107 T. Sugawa and Y. Miyajima: *IEEE Photonics Technol. Lett.* **3** (1991) 616.

108 Y. Miyajima, T. Komukai and T. Sugawa: *Electron. Lett.* **26** (1990) 194.

109 Y. Ohishi, T. Kanamori, J. Temmyo, M. Wada, M. Yamada, M. Shimizu, *et al.*: *Electron. Lett.* **27** (1991) 1995.

110 H. Yanagita, K. Itoh, E. Ishikawa, H. Aoki and H. Toratani: *OFC'95*, PD2-1.

111 Y. Ohishi, T. Kanamori, T. Kitagawa, S. Takahashi, E. Snitzer and G. H. Sigel, Jr.: *Opt. Lett.* **16** (1991) 1747.

112 Y. Durteste, M. Monerie, J. Y. Allain and H. Poignant: *Electron. Lett.* **27** (1991) 626.

113 S. F. Carter, D. Szebesta, S. T. Davey, R. Wyatt, M. C. Brierley and P. W. France: *Electron. Lett.* **27** (1991) 628.

114 M. Yamada, M. Shimizu, Y. Ohishi, J. Temmyo, M. Wada, T. Kanamori *et al.*: *IEEE Photonics Technol. Lett.* **4** (1992) 994.

115 S. Sanders, K. Dzurko, R. Parke, S. O'Brien, D. F. Welch, S. G. Grubb, *et al.*: *Electron. Lett.* **32** (1996) 343.

116 T. Sugawa, Y. Miyajima and T. Komukai: *Electron. Lett.* **26** (1990) 2042.

117 M. Shimizu, T. Kanamori, J. Temmyo, M. Wada, M. Yamada, T. Terunama, *et al.*: *IEEE Photonics Technol. Lett.* **5** (1993) 654.

118 Y. Miyajima, T. Sugawa and Y. Fukasaku: *Electron. Lett.* **27** (1991) 1706.

119 M. Yamada, M. Shimizu, H. Yoshinaga, K. Kikushima, T. Kanamori, Y. Ohishi, *et al.*: *Electron. Lett.* **31** (1995) 806.

120 H. Yamada, M. Shimizu, T. Kanamori, Y. Ohishi, Y. Terunuma, K. Oikawa, *et al.*: *IEEE Photonics Technol. Lett.* **7** (1995) 869.

121 M. Yamada, T. Kanamori, Y. Ohishi, M. Shimizu, Y. Terunuma, S. Sato, *et al.*: *IEEE Photonics Technol. Lett.* **9** (1997) 321.

122 Y. Nishida, T. Kanamori, Y. Ohishi, M. Yamada, K. Kobayashi and S. Sudo: *IEEE, Photonics Technol. Lett.* **9** (1997) 318.

123 H. Aoki, K. Itoh, E. Ishikawa, H. Yanagita and H. Toratani: *The Institute of Electronics, Information and Communication Engineers*. Technical Report (in Japanese) LQE96-28 (1996) p. 49.

124 E. Ishikawa, H. Yanagita, K. Itoh, H. Aoki and H. Toratani: *OSA Topical Meeting on Optical Amplifiers and Their Applications* (1996) Vol. 5, OSA Program Committee eds. (Optical Society of America, Washington, DC, 1996) p. 116.

125 H. Tawarayama, E. Ishikawa, K. Itoh, H. Aoki, H. Yanagita, K. Okada, *et al.*:

Topical Meeting, Optical Amplifiers and Their Applications, Victoria, (Optical Society of America, Washington, DC, 1997) PD1-1.

126 T. Shimada, Y. Nishida, K. Kobayashi, T. Kanamori, K. Oikawa, M. Yamada *et al.*: *Electron. Lett.* **33** (1997) 197.

127 M. Øbro, B. Pedersen, A. Bjarklev, J. H. Povlsen and J. E. Pedersen: *Electron. Lett.* **27** (1991) 470.

128 Y. Ohishi, T. Kanamori, T. Kitagawa, S. Takahashi, E. Snitzer and G. H. Sigel Jr: *Digest of Conference on Optical Fiber Communication* (Optical Society of America, Washinton, DC, 1991) PD2.

129 Y. Miyajima, T. Sugawa and Y. Fukasaku: *Tech. Dig. Optical Amplifiers and Their Applications*, Snowmass Village (1991) PD1.

130 B. Pedersen, W. J. Miniscalco and R. S. Quimby: *IEEE Photonics Technol. Lett.* **4** (1992) 446.

131 R. S. Quimby and B. Zheng: *Appl. Phys. Lett.* **60** (1992) 1055.

132 Y. Ohishi and E. Snitzer: *Proc. Int. Conf. Science and Technology of New Glasses*, eds. S. Sakka and N. Soga (The Ceramic Society of Japan, Tokyo, 1991) p. 199.

133 S. F. Carter, R. Wyatt, D. Szebesta and S. T. Davey: *17th European Conf. on Optical Communication and 8th Int. Conf. on Integrated Optics and Optical Fiber Communication*, Vol. 1 Part 1 (Paris, 1991) p. 21.

134 D. R. Simons, A. J. Faber, F. Simonis and H. De Waal: *8th Int. Symp. on Halide Glasses*. Perros-Guirec (1992) p. 448.

135 J. Y. Allain, M. Monerie and H. Poignant: *Electron. Lett.* **27** (1991) 1012.

136 J. Y. Allain, M. Monerie and H. Poignant: *Electron. Lett.* **28** (1992) 988.

137 M. Shimokozono, N. Sugimoto, A. Tate, Y. Katoh, M. Tanno, S. Fukuda, *et al.*: *Appl. Phys. Lett.* **68** (1996) 2177.

138 Y. Ohishi, T. Kanamori, T. Nishi, Y. Nishida and S. Takahashi: *Proc. 16th Int. Congress on Glass*, ed. D. E. de Ceramica y Vitrio, **2** (1992) p. 73.

139 K. Wei, D. P. Machewirth, J. Wenzel, E. Snitzer and G. H. Sigel, Jr.: *J. Non-Cryst. Solids* **182** (1995) 257.

140 B. Dussardier, D. W. Hewak, B. N. Samson, H. J. Tate, J. Wang and D. N. Payne: *Electron. Lett.* **31** (1995) 206.

141 L. R. Copeland, W. A. Reed, M. R. Shahriari, T. Iqbal, P. Hajcak and G. H. Sigel, Jr.: *Mat. Res. Soc. Symp. Proc.* **244** (1992) p. 203.

142 D. R. Simons, A. J. Faber and H. de Waal: *Opt. Lett.* **20** (1995) 468.

143 D. W. Hewak, J. A. Medeiros Neto, D. Samson, R. S. Brown, K. P. Jedrzejewski, J. Wang *et al.*: *IEEE Photonics Technol. Lett.* **6** (1994) 609.

144 Y. Ohishi, A. Mori, T. Kanamori, K. Fujiura and S. Sudo: *Appl. Phys. Lett.* **65** (1994) 13.

145 D. R. Simons, A. J. Faber and H. de Waal: *J. Non-Cryst. Solids* **185** (1995) 283.

146 J. S. Wang, E. M. Vogel, E. Snitzer, J. L. Jackel, V. L. da Silva and Y. Silberberg: *J. Non-Cryst. Solids* **178** (1994) 109.

147 E. R. Taylor, B. N. Samson, D. W. Hewak, J. A. Medeiros Neto, D. N. Payne, S. Jordery *et al.*: *J. Non-Cryst. Solids* **184** (1995) 61.

148 K. Kadono, H. Higuchi, M. Takahashi, Y. Kawamoto and H. Tanaka: *J. Non-Cryst. Solids* **184** (1995) 309.

149 D. Marchese, G. Kakarantzas and A. Jha: *J. Non-Cryst. Solids* **196** (1996) 314.

150 J. A. Medeiros Neto, E. R. Taylor, B. N. Samson, J. Wang, D. W. Hewak, R. I. Laming, *et al.*: *J. Non-Cryst. Solids* **184** (1995) 292.

151 P. C. Becker, M. M. Broer, V. C. Lambrecht, A. J. Bruce and G. Nykolak:

Technical Digest of Topical Meeting in Optical Amplifiers and Their Applications, (Optical Society of America, Washington, DC, 1992) PD5.

152 D. W. Hewak, R. S. Deol, J. Wang, G. Wylangowski, J. A. Medeiros Neto, B. N. Samson, *et al.*: *Electron. Lett.* **29** (1993) 237.

153 R. S. Deol. D. W. Hewak, S. Jorderg, A. Jha, M. Poulain, M. D. Baro, *et al.*: *J. Non-Cryst. Solids* **161** (1993) 257.

154 V. M. Krasteva, G. H. Sigel, Jr., S. L. Semjonov, M. M. Bubnov and M. I. Belovolov: *OSA Topical Meetings, Optical Amplifiers and Their Applications*, Victoria (Optical Society of America, Washington, DC, 1997) TuA4.

155 D. P. Machewirth, K. Wei, V. Krasteva, R. Datta, E. Snitzer and G. H. Sigel, Jr.: *J. Non-Cryst. Solids* **213 & 214** (1997) 295.

156 J. Wang, J. R. Hector, D. Brady, D. Hewak, B. Brocklesby, M. Kluth, *et al.*: *Appl. Phys. Lett.* **71** (1997) 1753.

157 Y. G. Choi and J. Heo: *J. Non-Cryst. Solids* **217** (1997) 199.

158 K. Abe, H. Takebe and K. Morinaga: *J. Non-Cryst. Solids* **212** (1997) 143.

159 H. Takebe, K. Yoshino, T. Murata, K. Morinaga, J. Hector, W. S. Brocklesby, *et al.*: *Appl. Opt.* **36** (1997) 5839.

160 T. Izumitani, T. Yamashita, M. Tokida, K. Miura and M. Tajima: *Materials Science Forum*, **19–20** (Trans. Tech. Publication Ltd, Switzerland, 1987) p. 19.

161 E. Ishikawa, H. Yanagita, H. Tawarayama, K. Itoh, H. Aoki, K. Yamanaka, *et al.*: *Technical Digest, Optical Amplifiers and Their Applications*, July 27–29, Vail, Colorado (Optical Society of America, Washington, DC, 1998) p. 216.

162 D. W. Hewak, B. N. Samson, J. A. Medeiros Neto, R. I. Laming and D. N. Payne: *Electron. Lett.* **30** (1994) 968.

163 K. Wei, D. P. Machewirth, J. Wenzel, E. Snitzer and G. H. Sigel Jr.: *Opt. Lett.* **19** (1994) 904.

164 J. Heo and Y. B. Shin: *J. Non-Cryst. Solids* **196** (1996) 162.

165 B. N. Samson, J. A. Medeiros Neto, R. I. Laming and D. W. Hewak: *Electron. Lett.* **30** (1994) 1617.

166 S. Tanabe, T. Hanada, M. Watanabe, T. Hayashi and N. Soga: *J. Am. Ceram. Soc.* **78** (1995) 2917.

167 M. Shimizu, H. Suda and M. Horiguchi: *Electron. Lett.* **23** (1987) 768.

168 Y. Ohishi, T. Kanamori and S. Takahashi: *IEEE Photonics Technol. Lett.* **3** (1991) 688.

169 J. Y. Allain, M. Monerie and H. Poignant: *Electron. Lett.* **27** (1991) 189.

170 R. G. Smart, D. C. Hanna. A. C. Tropper, S. T. Davey, S. F. Carter and D. Szebesta: *Electron. Lett.* **27** (1991) 1307.

171 R. G. Smart, J. N. Carter, A. C. Tropper. D. C. Hanna. S. T. Davey, S. F. Carter, *et al.*: *Opt. Commun.* **86** (1991) 333.

172 Y. Ohishi and T. Kanamori: *Electron. Lett.* **28** (1992) 162.

173 T. Sugawa, E. Yoshida, Y. Miyajima and M. Nakazawa: *Electron. Lett.* **29** (1993) 902.

174 Y. Shi, C. V. Poulsen, M. Sejka, M. Ibsen and O. Poulsen: *Electron. Lett.* **29** (1993) 1426.

175 Y. Zhao and S. Poole: *Electron. Lett.* **30** (1994) 967.

176 Y. Zhao, S. Fleming and S. Poole: *Opt. Commun.* **114** (1995) 285.

177 H. Döring, J. Peupelmann and F. Wenzl: *Electron. Lett.* **31** (1995) 1068.

178 J. Y. Allain, M. Monerie and H. Poignant: *Electron. Lett.* **27** (1991) 1156.

179 D. Piehler, D. Craven, N. Kwong and H. Zarem: *Electron. Lett.* **29** (1993) 1857.

180 D. M. Baney, L. Yang, J. Ratcliff and Kok Wai Chang: *Electron. Lett.* **31** (1995) 1842.

181 L. Reekie, R. J. Mears, S. B. Poole and D. N. Payne: *J. Lightwave Technol.* **LT-4** (1986) 956.

182 M. C. Brierley and P. W. France: *Electron. Lett.* **23** (1987) 815.

183 L. Reekie, I. M. Jauncey, S. B. Poole and D. N. Payne: *Electron. Lett.* **23** (1987) 884.

184 W. J. Miniscalco, L. J. Andrews, B. A. Thompson, R. S. Quimby, L. J. B. Vacha and M. G. Drexhage: *Electron. Lett.* **24** (1988) 28.

185 M. C. Brierley and C. A. Millar: *Electron. Lett.* **24** (1988) 438.

186 K. Liu, M. Digonnet, K. Fesler, B. Y. Kim and H. J. Show: *Electron. Lett.* **24** (1988) 838.

187 F. Hakimi, H. Po, R. Tumminelli, B. C. McCollum, L. Zenteno, N. M. Cho, *et al.*: *Opt. Lett.* **14** (1989) 1060.

188 L. Wetenkamp, Ch. Frerichs, G. F. West and H. Többen: *J. Non-Cryst. Solids* **140** (1992) 19.

189 T. Komukai, Y. Fukasaku, T. Sugawa and Y. Miyajima: *Electron. Lett.* **29** (1993) 755.

190 D. S. Funk, J. W. Carlson and J. G. Eden: *Electron. Lett.* **30** (1994) 1859.

191 J. Wang, L. Reekie, W. S. Brocklesby, Y. T. Chow and D. N. Payne: *J. Non-Cryst. Solids* **180** (1995) 207.

192 M. C. Farries, P. R. Morkel and J. E. Townsend: *Electron. Lett.* **24** (1988) 709.

193 M. C. Brieley, P. W. France and C. A. Millar: *Electron. Lett.* **24** (1988) 539.

194 D. C. Hanna, R. M. Percival, R. G. Smart, J. E. Townsend and A. C. Tropper: *Electron. Lett.* **25** (1989) 593.

195 L. Wetenkamp: *Electron. Lett.* **26** (1990) 883.

196 J. Y. Allain, M. Monerie and H. Poignant: *Electron. Lett.* **26** (1990) 261.

197 H. Többen: *Summaries of Advanced Solid-State Lasers Topical Meeting*, Santa Fe (The Optical Society of America, Washington, DC, 1992) p. 63.

198 J. Schneider: *Electron. Lett.* **31** (1995) 1250.

199 J. Y. Allain, M. Monerie and H. Poignant: *Electron. Lett.* **27** (1991) 1513.

200 R. M. Percival, D. Szebesta, S. T. Davey, N. A. Swain and T. A. King: *Electron. Lett.* **28** (1992) 2064.

201 R. M. Percival, D. Szebesta, S. T. Davey, N. A. Swain and T. A. King: *Electron. Lett.* **28** (1992) 2231.

202 M. C. Brierley and P. W. France: *Electron. Lett.* **24** (1988) 935.

203 R. J. Mears, L. Reekie, S. B. Poole and D. N. Payne: *Electron. Lett.* **22** (1986) 159.

204 I. M. Jauncey, L. Reekie, R. J. Mears and C. J. Rowe: *Opt. Lett.* **12** (1987) 164.

205 C. A. Millar, I. D. Miller, B. J. Ainslie, S. P. Craig and J. R. Armitage: *Electron. Lett.* **23** (1987) 865.

206 L. Reekie, I. M. Jauncey, S. B. Poole and D. N. Payne: *Electron. Lett.* **23** (1987) 1076.

207 Y. Kimura and M. Nakazawa: *J. Appl. Phys.* **64** (1988) 516.

208 J. Y. Allain, M. Monerie and H. Poignant: *Electron. Lett.* **25** (1989) 28.

209 M. S. O'Sullivan, J. Chrostowski, E. Desurvire and J. R. Simpson: *Opt. Lett.* **14** (1989) 438.

210 P. L. Scrivener, E. J. Tarbox and P. D. Maton: *Electron. Lett.* **25** (1989) 549.

211 J. Y. Allain, M. Monerie and H. Poignant: *Electron. Lett.* **25** (1989) 1082.

212 K. Suzuki, Y. Kimura and M. Nakazawa: *Jpn. J. Appl. Phys.* **28** (1989) L1000.

213 R. G. Smart, J. N. Carter, D. C. Hanna and A. C. Tropper: *Electron. Lett.* **26** (1990) 649.
214 H. Yanagita, I. Masuda, T. Yamashita and H. Toratani: *Electron. Lett.* **26** (1990) 1836.
215 T. J. Whitley, C. A. Millar, R. Wyatt, M. C. Brieley and D. Szebesta: *Electron. Lett.* **27** (1991) 1785.
216 J. Y. Allain, M. Monerie and H. Poignant: *Electron. Lett.* **28** (1992) 111.
217 H. Többen: *Electron. Lett.* **28** (1992) 1361.
218 J. F. Massicott, M. C. Brierley, R. Wyatt, S. T. Davey and D. Szebesta: *Electron. Lett.* **29** (1993) 2119.
219 C. Ghisler, M. Pollnau, G. Burea, M. Bunea, W. Lüthy and H. P. Weber: *Electron. Lett.* **31** (1995) 373.
220 D. C. Hanna, R. M. Percival, I. R. Perry, R. G. Smart and A. C. Tropper: *Electron. Lett.* **24** (1988) 1068.
221 G. T. Maker and A. I. Ferguson: *Electron. Lett.* **24** (1988) 1160.
222 M. E. Fermann, D. C. Hanna, D. P. Shepherd, P. J. Suni and J. E. Townsend: *Electron. Lett.* **24** (1988) 1135.
223 J. T. Kringlebotn, J.-L. Archambault, L. Reekie, J. E. Townsend, G. G. Vienne and D. N. Payne: *Electron. Lett.* **30** (1994) 972.
224 G. P. Lees, A. Hartog, A. Leach and T. P. Newson: *Electron. Lett.* **31** (1995) 1836.
225 L. Esterowitz, R. Allen and I. Aggarwal: *Electron. Lett.* **24** (1988) 1104.
226 D. C. Hanna, I. M. Jauncey, R. M. Percival, I. R. Perry, R. G. Smart, P. J. Suni, *et al.*: *Electron. Lett.* **24** (1988) 1222.
227 R. Allen and L. Esterowitz: *Appl. Phys. Lett.* **55** (1989) 721.
228 D. C. Hanna, M. J. McCarthy, I. R. Perry and P. J. Suni: *Electron. Lett.* **25** (1989) 1365.
229 J. Y. Allain, M. Monerie and H. Poingant: *Electron. Lett.* **25** (1989) 1660.
230 D. C. Hanna, I. R. Perry, J. R. Lincoln and J. E. Townsend: *Opt. Commun.* **80** (1990) 52.
231 J. Y. Allain, M. Monerie and H. Poignant: *Electron. Lett.* **26** (1990) 166.
232 D. C. Hanna, R. M. Percival, R. G. Smart and A. C. Tropper: *Opt. Commun.* **75** (1990) 283.
233 J. N. Carter, R. G. Smart, D. C. Hanna and A. C. Tropper: *Electron. Lett.* **26** (1990) 599.
234 W. L. Barnes and J. E. Townsend: *Electron. Lett.* **26** (1990) 746.
235 J. N. Carter. R. G. Smart, D. C. Hanna and A. C. Tropper: *Electron. Lett.* **26** (1990) 1759.
236 R. G. Smart, J. N. Carter, A. C. Tropper and D. C. Hanna: *Opt. Commun.* **82** (1991) 563.
237 R. M. Percival, S. F. Carter, D. Szebesta, S. T. Davey and W. A. Stallard: *Electron. Lett.* **27** (1991) 1912.
238 R. M. Percival, D. Szebesta and S. T. Davey: *Electron. Lett.* **28** (1992) 671.
239 T. Komukai, T. Yamamoto, T. Sugawa and Y. Miyajima: *Electron. Lett.* **28** (1992) 830.
240 J. R. Lincoln, C. J. Mackechnie, J. Wang, W. S. Brocklesby, R. S. Deol, A. Pearson *et al.*: *Electron. Lett.* **28** (1992) 1021.
241 S. G. Grubb, K. W. Bennett, R. S. Cannon and W. F. Humer: *Electron. Lett.* **28** (1992) 1243.
242 C. J. Mackechnie, W. L. Barnes, D. C. Hanna and J. E. Townsend: *Electron. Lett.* **29** (1993) 52.

243 Y. Miyajima, T. Komukai and T. Sugawa: *Electron. Lett.* **29** (1993) 660.

244 T. Yamamoto, Y. Miyajima, T. Komukai and T. Sugawa: *Electron. Lett.* **29** (1993) 986.

245 R. M. Percival, D. Szebesta, C. P. Seltzer, S. D. Perrin, S. T. Davey and M. Louka: *Electron. Lett.* **29** (1993) 2110.

246 M. L. Dennis, J. W. Dixon and I. Aggarwal: *Electron. Lett.* **30** (1994) 136.

247 T. Yamamoto, Y. Miyajima and T. Komukai: *Electron. Lett.* **30** (1994) 220.

248 R. M. Percival, D. Szebesta and J. R. Williams: *Electron. Lett.* **30** (1994) 1057.

249 R. M. Percival, D. Szebesta, C. P. Seltzer, S. D. Perrin, S. T. Davey and M. Louka: *IEEE, J. Quantum Electron.* **QE 31** (1995) 489.

250 G. Tohmon, J. Ohya, H. Sato and T. Uno: *IEEE Photonics. Technol. Lett.* **7** (1995) 742.

251 F. J. McAleavey and B. D. MacCraith: *Electron. Lett.* **31** (1995) 800.

252 S. Sanders, R. G. Waarts, D. G. Mehuys and D. F. Welch: *Appl. Phys. Lett.* **67** (1995) 1815.

253 D. C. Hanna, R. M. Percival, I. R. Perry, R. G. Smart, P. J. Suni, J. E. Townsend, *et al.*: *Electron. Lett.* **24** (1988) 1111.

254 J. R. Armitage, R. Wyatt, B. J. Ainslie and S. P. Craig-Ryan: *Electron. Lett.* **25** (1989) 298.

255 J. Y. Allain, J. F. Bayon, M. Monerie, P. Bernage and P. Niay: *Electron. Lett.* **29** (1993) 309.

256 H. M. Pask, J. L. Archambault, D. C. Hanna, L. Reekie, P. St. J. Russel, J. E. Townsend, *et al.*: *Electron. Lett.* **30** (1994) 863.

257 D. S. Funk, S. B. Stevens, S. S. Wu and J. G. Eden: *IEEE J. Quantum Electron.* **QE32** (1996) 638.

258 L. M. Yang, D. T. Walton, J. Nees and W. H. Weber: *Electron. Lett.* **32** (1996) 658.

259 R. Wyatt, B. J. Ainslie and S. P. Craig: *Electron. Lett.* **24** (1988) 1362.

260 W. L. Barnes, S. B. Poole, J. E. Townsend, L. Reekie, D. J. Taylor and D. N. Payne: *J. Lightwave Technol.* **7** (1989) 1461.

261 A. B. Grudinin, D. J. Richardson, A. K. Senatorov and D. N. Payne: *Electron. Lett.* **28** (1992) 766.

262 H. Többen: *Electron. Lett.* **29** (1993) 667.

263 Ch. Frerichs and T. Tauermann: *Electron. Lett.* **30** (1994) 706.

264 I. M. Jauncey, L. Reekie, J. E. Townsend, D. N. Payne and C. J. Rowe: *Electron. Lett.* **24** (1988) 24.

265 D. M. Pataca, M. L. Rocha, K. Smith, T. J. Whitley and R. Wyatt: *Electron. Lett.* **30** (1994) 964.

266 T. Komukai, T. Yamamoto, T. Sugawa and Y. Miyajima: *IEEE J. Quantum Electron.* **QE31** (1995) 1880.

267 I. J. Booth, C. J. Mackechnie and B. F. Ventrudo: *IEEE J. Quantum Electron.* **QE32** (1996) 118.

268 R. M. Percival, D. Szebesta and S. T. Davey: *Electron. Lett.* **28** (1992) 1866.

269 R. M. Percival, D. Szebesta and S. T. Davey: *Electron. Lett.* **29** (1993) 1054.

270 H. M. Pask, R. J. Carman, D. C. Hanna, A. C. Tropper, C. J. Mackechnie, P. R. Barber, *et al.*: *IEEE J. Selected Topics in Quantum Electron.* **1** (1995) 2.

271 D. S. Funk, J. W. Carlson and J. G. Eden: *OSA Proceedings on Advanced Solid State Lasers* **24**, ed. B. H. T. Chai and S. A. Payne (Optical Society of America, Washington, DC, 1995) p. 71.

272 M. Wegmuler, W. Hodel and H. P. Weber: *Opt. Commun.* **127** (1996) 266.

273 Y. Zhao and S. Fleming: *Electron. Lett.* **32** (1996) 1199.
274 D. M. Baney, G. Rankin and Kok-Wai Chang: *Opt. Lett.* **21** (1996) 1372.
275 H. M. Pask, A. C. Tropper and D. C. Hanna: *Opt. Commun.* **134** (1997) 139.
276 D. M. Baney, G. Rankin and Kok Wai Chang: *Appl. Phys. Lett.* **69** (1996) 1662.
277 A. Mori, Y. Ohishi and S. Sudo: *Electron. Lett.* **33** (1997) 863.
278 T. Schweizer, D. W. Hewak, D. N. Payne, T. Jensen and G. Huber: *Electron. Lett.* **32** (1996) 666.
279 G. Tohmon, H. Sato, J. Ohya and T. Uno: *Appl. Opt.* **36** (1997) 3381.
280 J. Schneider, C. Carbonnier and U. B. Unrau: *Appl. Opt.* **36** (1997) 8595.
281 D. S. Funk, J. G. Eden, J. S. Osinski and B. Lu: *Electron. Lett.* **33** (1997) 1958.
282 I. P. Alcock, A. I. Ferguson, D. C. Hanna and A. C. Tropper: *Opt. Commun.* **58** (1986) 405.
283 C. Carbonnier, H. Többen and U. B. Unrau: *Electron. Lett.* **34** (1998) 893.
284 M. J. Digonnet: *J. Lightwave Technol.* **LT-4** (1986) 1631.
285 K. Liu, M. Digonnet, H. J. Show, B. J. Ainslie and S. P. Craig: *Electron. Lett.* **23** (1987) 1320.
286 K. A. Fesler, B. Y. Kim and H. J. Show: *Electron. Lett.* **25** (1989) 534.
287 C. B. Morrison, L. M. Zinkiewicz, J. Niesen and L. Figueroa: *SPIE* **566**, Fiber Optic and Laser Sensors III, (1985) p. 99.
288 K. A. Fesler, M. J. F. Digonnet, B. Y. Kim and H. J. Show: *Opt. Lett.* **15** (1990) 1321.
289 D. C. Hanna, I. R. Perry, R. G. Smart, P. J. Suni, J. E. Townsend and A. C. Tropper: *Opt. Commun.* **72** (1989) 230.
290 Y. Ohishi, T. Kanamori and S. Takahashi: *Jpn. J. Appl. Phys.* **30** (1991) L1282.
291 K. Iwatsuki: *IEEE Photonic Technol. Lett.* **2** (1991) 237.
292 L. A. Wang and C. D. Chen: *Electron. Lett.* **33** (1997) 703.
293 D. N. Wang, B. T. Meggitt, A. W. Palmer, K. T. V. Grattan and Y. N. Ning: *IEEE Photonics Technol. Lett.* **7** (1995) 620.
294 S. Gray, J. D. Minelly, A. B. Grudinin and J. E. Caplen: *Electron. Lett.* **33** (1997) 1382.
295 F. Auzel, D. Meichenin and H. Poignant: *Electron. Lett.* **24** (1988) 909.
296 C. Ghisler, W. Lüthy, H. P. Weber: *IEEE J. Quantum Electron.* **31** (1995) 1877.
297 F. Chenard: *Photonics Spectra* (1994, May) p. 126.
298 M. Nakazawa, M. Tokuda, K. Washio and Y. Asahara: *Opt. Lett.* **9** (1984) 312.
299 M. J. Weber, J. E. Lynch, D. H. Blackburn and D. J. Cronin: *IEEE Quantum Electron.* **QE-19** (1983) 1600.
300 M. J. Weber and J. E. Lynch: *1982 Laser Program Annual Report*, (Lawrence Livermore National Laboratory, 1983) p. 7-45 and refs. cited therein.
301 A. Mori, Y. Ohishi, T. Kanamori and S. Sudo: *Appl. Phys. Lett.* **70** (1997) 1230.
302 H. Higuchi, M. Takahashi, Y. Kawamoto, K. Kadono, T. Ohtsuki, N. Peyghambarian and N. Kitamura: *J. Appl. Phys.* **83** (1998) 19.
303 S. Q. Gu, S. Ramachandran, E. E. Reuter, D. A. Turnball, J. T. Verdeyen and S. G. Bishop: *J. Appl. Phys.* **77** (1995) 3365.
304 S. Q. Gu, S. Ramachandran, E. E. Reuter, D. A. Turnball, J. T. Verdeyen and S. G. Bishop: *Appl. Phys. Lett.* **66** (1995) 670.
305 C. C. Ye, M. Hempstead, D. W. Hewak and D. N. Payne: *IEEE Photonics Technol. Lett.* **9** (1997) 1104.
306 J. R. Hector, D. W. Hewak, J. Wang, R. C. Moore and W. S. Brocklesby: *IEEE, Photonics Technol. Lett.* **9** (1997) 443.
307 Y. S. Kim, W. Y. Cho, Y. B. Shin and J. Heo: *J. Non-Cryst. Solids* **203** (1996) 176.

308 P. A. Tick, N. F. Borrelli, L. K. Cornelius and M. A. Newhouse: *J. Appl. Phys.* **78** (1995) 6367.

309 R. S. Quimby, P. A. Tick, N. F. Borrelli and L. K. Cornelius: *Proc. CLEO'97* (1997) p. 79.

310 R. S. Quimby, P. A. Tick, N. F. Borrelli and L. K. Cornelius: *J. Appl. Phys.* **83** (1998) 1649.

311 J. Qiu, N. Sugimoto and K. Hirao: *J. Mat. Sci. Lett.* **15** (1996) 1641.

312 Y. Wang and J. Ohwaki: *Appl. Phys. Lett.* **63** (1993) 3268.

313 Y. Kawamoto, R. Kanno and J. Qiu: *J. Mat. Sci.* **33** (1998) 63.

314 K. Hirao, K. Tanaka, M. Makita and N. Soga: *J. Appl. Phys.* **78** (1995) 3445.

315 E. Snitzer, H. Po, F. Hakimi, R. Tumminelli and B. C. McCollum: *Proc. Optical Fiber Sensor, OSA Technical Digest Series*, **2** (Optical Society of America, Washington, DC. 1988) PD5.

316 H. Po, E. Snitzer, R. Tumminelli, L. Zenteno, F. Hakimi, N. M. Cho and T. Haw: *OFC'89*, Post-deadline paper PD-7 (1989).

317 H. Po, J. D. Cao, B. M. Laliberte, R. A. Minns and R. F. Robinson: *Electron. Lett.* **29** (1993) 1500.

318 E. Chen, A. B. Grudinin, J. Porta and J. D. Minelly: *CLEO'98*, Technical Digest Series **6** (1998) p. 78.

319 Z. J. Chen, A. B. Grudinin, J. Porta and J. D. Minelly: *Opt. Lett.* **23** (1998) 454.

320 M. Muendel, B. Engstrom, D. Kea, B. Laliberte, R. Minns, R. Robinson, *et al.*: *CLEO'97*, Post-deadline papers CPD30-1 (1997).

321 E. Lallier: *Appl. Opt.* **31** (1992) 5276.

322 T. Izawa and H. Nakagome: *Appl. Phys. Lett.* **21** (1972) 584.

323 R. V. Ramaswamy and R. Srivastava: *J. Lightwave Technol.* **6** (1988) 984.

324 J. Albert: *Introduction to Glass Integrated Optics*, ed. S. I. Najafi (Artech House, Inc., Boston, 1992) p.7.

325 L. Roß: *Glasstech. Ber.* **62** (1989) 285.

326 R. Jansen: *Optical Fiber Communication Conference (OFC'90)*, Technical Digest Series **1** (Optical Society of America, Washington, DC, 1990) p. 62.

327 N. A. Sanford, K. J. Malone and D. R. Larson: *Opt. Lett.* **15** (1990) 366.

328 E. K. Mwarania, L. Reekie, J. Wang and J. S. Wilkinson: *Electron. Lett.* **26** (1990) 1317.

329 S. I. Najafi, W. J. Wang, J. F. Currie, R. Leonelli and J. L. Brebner: *IEEE Photonics Technol. Lett.* **1** (1989) 109.

330 S. I. Najafi, W. Wang, J. F. Currie, R. Leonelli and J. L. Brebner: *SPIE* **1128** Glass for Optoelectronics (1989) p. 42.

331 H. Aoki, O. Maruyama and Y. Asahara: *Optical Fiber Communication Conference (OFC'90)*, Technical Digest Series **1** (Optical Society of America, Washington, DC, 1990) p. 201.

332 H. Aoki, O. Maruyama and Y. Asahara: *IEEE Photonics Technol. Lett.* **2** (1990) 459.

333 M. Kawachi: *Opt. Quantum Electron.* **22** (1990) 391.

334 Y. Hibino, T. Kitagawa, M. Shimizu, F. Hanawa and A. Sugita: *IEEE Photonics Technol. Lett.* **1** (1989) 349.

335 T. Kitagawa, K. Hattori, M. Shimizu, Y. Ohmori and M. Kobayashi: *Electron. Lett.* **27** (1991) 334.

336 R. Tumminelli, F. Hakimi and J. Haavisto: *Opt. Lett.* **16** (1991) 1098.

337 H. Yajima, S. Kawase and Y. Sekimoto: *Appl. Phys. Lett.* **21** (1972) 407.

338 B.-U. Chen and C. L. Tang: *Appl. Phys. Lett.* **28** (1976) 435.

339 G. R. J. Robertson and P. E. Jessop: *Appl. Opt.* **30** (1991) 276.

340 J. Shmulovich, A. Wong, Y. H. Wong, P. C. Becker, A. J. Bruce and R. Adar: *Electron. Lett.* **28** (1992) 1181.

341 M. Nakazawa and Y. Kimura: *Electron. Lett.* **28** (1992) 2054.

342 D. S. Gill, R. W. Eason, C. Zaldo, H. N. Rutt, N. B. Vainos: *J. Non-Cryst. Solids* **191** (1995) 321.

343 B. Wu and P. L. Chu: *IEEE Photonics Technol. Lett.* **7** (1995) 655.

344 D. P. Shephered, D. J. B. Brinck, J. Wang, A. C. Tropper, D. C. Hanna, G. Kakarantzas, *et al.*: *Opt. Lett.* **19** (1994) 954.

345 E. Snoeks, G. N. van den Hoven, A. Polman, B. Hendriksen, M. B. J. Diemeer and F. Priolo: *J. Opt. Soc. Am.* **B12** (1995) 1468.

346 M. Benatsou, B. Capoen, M. Bouazaoui, W. Tchana and J. P. Vilcot: *Appl. Phys. Lett.* **71** (1997) 428.

347 E. Fogret, G. Fonteneau, J. Lucas and R. Rimet: *Proc XVII Int. Congress on Glass* Vol. **5**, Special Glass (Chinese Ceramic Society, Beijing, 1995) p. 175.

348 C. Charron, E. Fogret, G. Fonteneau, R. Rimet and J. Lucas: *J. Non-Cryst. Solids* **184** (1995) 222.

349 E. Fogret, G. Fonteneau, J. Lucas and R. Rimet: *Opt. Mat.* **5** (1996) 79.

350 J. Lucas, X. H. Zhang, K. Le Foulgoc, G. Fonteneau and E. Fogret: *J. Non-Cryst. Solids* **203** (1996) 127.

351 E. Fogret, G. Fonteneau, J. Lucas and R. Rimet: *Opt. Mat.* **5** (1996) 87.

352 E. Josse, G. Fonteneau and J. Lucas: *Mat. Res. Bull.* **32** (1997) 1139.

353 R. P. de Melo, Jr., B. J. P. da Silva, E. L. Falcão-Filho, E. F. da Silva, Jr., D. V. Petrov, Cid B. de Araujo, *et al.*: *Appl. Phys. Lett.* **67** (1995) 886.

354 O. Perrot, L. Guinvarc'h, D. Benhaddou, P. C. Montomery, R. Rimet, B. Boulard, *et al.*: *J. Non-Cryst. Solids* **184** (1995) 257.

355 J. Ballato, R. E. Riman and E. Snitzer: *J. Non-Cryst. Solids* **213 & 214** (1997) 126.

356 M. Saruwatari and T. Izawa: *Appl. Phys. Lett.* **24** (1974) 603.

357 K. Hattori, T. Kitagawa, Y. Ohmori and M. Kobayashi: *IEEE Photonics Technol. Lett.* **3** (1991) 882.

358 J. E. Roman and K. A. Winick: *Technical Digest, Optical Society of America, 1992 Annual Meeting*, Albuquerque (Optical Society of America, Washington, DC, 1992) p. 58.

359 J. E. Roman and K. A. Winick: *Appl. Phys. Lett.* **61** (1992) 2744.

360 H. Aoki, O. Maruyama and Y. Asahara: *Electron. Lett.* **26** (1990) 1910.

361 K. J. Malone, N. A. Sanford and J. S. Hayden: *Electron. Lett.* **29** (1993) 691.

362 N. A. Sanford, K. J. Malone and D. R. Larson: *Opt. Lett.* **16** (1991) 1095.

363 J. R. Bonar, J. A. Bebbington, J. S. Aitchison, G. D. Maxwell and B. J. Ainslie: *Electron. Lett.* **30** (1994) 229.

364 T. Kitagawa, K. Hattori, Y. Hibino, Y. Ohmori and M. Horiguchi: *Proc. 18th European Conference on Optical Communication*, **3** (Berlin, 1992), ThPD II-5.

365 J. A. Aust, K. J. Malone, D. L. Veasey, N. A. Sanford and A. Roshko: *Opt. Lett.* **19** (1994) 1849.

366 J. Bonar, J. A. Bebbington, J. S. Aitchison, G. D. Maxwell and B. J. Ainslie: *Electron. Lett.* **31** (1995) 99.

367 V. François, T. Ohtsuki, N. Peyghambarian and S. I. Najafi: *Opt. Commun.* **119** (1995) 104.

368 T. Feuchter, E. K. Mwarania, J. Wang, L. Reekie and J. S. Wilkinson: *IEEE Photonics Technol. Lett.* **4** (1992) 542.

369 T. Kitagawa, F. Bilodeau, B. Malo, S. Theriault, J. Albert, D. C. Johnson, *et al.*:
     Electron. Lett. **30** (1994) 1311.
370 S. Kawanishi, K. Hattori, H. Takara, M. Oguma, O. Kamatani and Y. Hibino:
     Electron. Lett. **31** (1995) 363.
371 J. E. Roman, P. Camy, M. Hampstead, W. S. Brocklesby, S. Nouh, A. Beguin,
     et al.: *Electron. Lett.* **31** (1995) 1345.
372 G. L. Vossler, C. J. Brooks and K. A. Winick: *Electron. Lett.* **31** (1995) 1162.
373 A. Yeniay, J.-M. P. Delavauz, J. Toulouse, D. Barbier, T. A. Strasser and J. R.
     Pedrazanni: *Electron. Lett.* **33** (1997) 1792.
374 N. A. Sanford, K. J. Malone, D. R. Larson and R. K. Hickernell: *Opt. Lett.* **16**
     (1991) 1168.
375 T. Kitagawa, K. Hattori, Y. Hibino and Y. Ohmori: *J. Lightwave Technol.* **12**
     (1994) 436.
376 A. Yeniay, J.-M. P. Delavaux, J. Toulouse, D. Barbier, T. A. Strasser and J. R.
     Pedrazanni: *IEEE Photonics Technol. Lett.* **9** (1997) 1099.
377 A. Yeniay, J.-M. P. Delavaux, J. Toulouse, D. Barbier, T. A. Strasser, J. R.
     Pedrazzani, *et al.*: *IEEE Photonics Technol. Lett.* **9** (1997) 1580.
378 D. L. Veasey, D. S. Funk, N. A. Sanford and J. S. Hayden, *Appl. Phys. Lett.* **74**
     (1999) 789.
379 H. Aoki, E. Ishikawa and Y. Asahara: *Electron. Lett.* **27** (1991) 2351.
380 A. N. Miliou, X. F. Cao, R. Srivastava and R. V. Ramaswamy: *Technical Digest,
     Optical Society of America, 1992 Annual Meeting*, Albuquerque (Optical
     Society of America, Washington, DC, 1992) p. 57.
381 A. N. Miliou, X. F. Cao, R. Srivastava and R. V. Ramaswamy: *IEEE Photonics
     Technol. Lett.* **4** (1993) 416.
382 T. Kitagawa, K. Hattori, K. Shuto, M. Yasu, M. Kobayashi and M. Horiguchi:
     Electron. Lett. **28** (1992) 1818.
383 W. J. Wang, S. I. Najafi, S. Honkanen, Q. He, C. Wu and J. Glinski: *Electron. Lett.*
     28 (1992) 1872.
384 D. Trouchet, A. Beguin, P. Laborde and C. Lerminiaux: *Technical Digest Series
     4, Conference on Optical Fiber Communication/International Conference on
     Integrated Optics and Optical Fiber Communication*, San Jose (Optical Society
     of America, Washington, DC, 1993) p. 106.
385 J. Shmulovich, Y. H. Wong, G. Nykolak, P. C. Becker, R. Adar, A. J. Bruce, *et al.*:
     *Optical Fiber Conference (OFC)/Integrated Optics and Optical Fiber Commu-
     nication (IOOC)'93*, San Jose (Optical Society of America, Washington, DC,
     1993) PD18.
386 K. Hattori, T. Kitagawa, M. Oguma, M. Wada, J. Temmyo and M. Horiguchi:
     Electron. Lett. **29** (1993) 357.
387 R. N. Ghosh, J. Shmulovich, C. F. Kane, M. R. X. de Barros, G. Nykolak, A. J.
     Bruce *et al.*: *IEEE Photonics Technol. Lett.* **8** (1996) 518.
388 G. Nykolag, M. Haner, P. C. Becker, J. Shmulovich and Y. H. Wong: *IEEE
     Photonics Technol. Lett.* **5** (1993) 1185.
389 Y. C. Yan, A. J. Faber, H. de Waal, P. G. Kik and A. Polman: *Appl. Phys. Lett.* **71**
     (1997) 2922.
390 P. Camy, J. E. Roman, F. W. Willems, M. Hempstead, J. C. van der Plaats, C. Prel,
     et al.: *Electron. Lett.* **32** (1996) 321.
391 P. Fournier, P. Meshkinfam, M. A. Fardad, M. P. Andrews and S. I. Najafi:
     Electron. Lett. **33** (1997) 293.
392 T. Kitagawa, K. Hattori, K. Shuto, M. Yasu, M. Kobayashi and M. Horiguchi:

Technical Digest, Optical Amplifiers and Their Applications, Topical Meeting, Santa Fe (1992) PD1.

393 O. Lumholt, A. Bjarklev, T. Rasmussen and C. Lester: *J. Lightwave Technol.* **13** (1995) 275.

394 E. Snoeks, G. N. van den Hoven and A. Polman: *IEEE J. Quantum Electron*, **32** (1996) 1680.

395 K. Hattori, T. Kitagawa, M. Oguma, H. Okazaki and Y. Ohmori: *J. Appl. Phys.* **80** (1996) 5301.

396 T. Ohtsuki, S. Honkanen, S. I. Najafi and N. Peyghambarian: *J. Opt. Soc. Am.* **B14** (1997) 1838.

397 F. Di Pasquale and M. Federighi: *IEEE J. Quantum Electron.* **30** (1994) 2127.

398 M. Federighi and F. Di Pasquale: *IEEE Photonics Technol. Lett.* **7** (1995) 303.

399 D. Barbier, M. Rattay, F. Saint Andre, G. Clauss, M. Trouillon, A. Kevorkian, *et al.*: *IEEE Photonics Technol. Lett.* **9** (1997) 315.

400 E. K. Mwarania, D. M. Murphy, M. Hempstead, L. Reekie and J. S. Wilkinson: *IEEE Photonics Technol. Lett.* **4** (1992) 235.

401 M. Oguma, T. Kitagawa, K. Hattori and M. Horiguchi: *IEEE Photonics Technol. Lett.* **6** (1994) 586.

4

Nonlinear optical glass

Introduction

There has been considerable activity in the development of photonics devices such as all-optical switches and modulators for many applications in optical communication and optical computer systems. The success of such devices depends on the development of nonlinear optical materials. Various types of glass are very attractive materials for these applications, because of their strong nonlinearities and their relatively fast response time, and in addition, they are mechanically durable and are compatible with fiber and waveguide fabrication procedures. They are also very attractive from the basic viewpoint of understanding the physical mechanisms of optical nonlinearities in materials. Generally, optical nonlinearities of glass materials can be divided into two principal categories [1, 2]: resonant and nonresonant.

Nonresonant nonlinear glasses range from optical glasses to high-refractive-index glasses such as heavy-metal oxide glasses and chalcogenide glasses, and resonant nonlinear glasses including composite glasses containing semiconductor or metal microcrystallites. These have been the subject of investigation, although further developments in photonics switching and information processing will depend critically on the continuing development of improved glass materials. In this chapter, we first give a brief survey of the fundamentals of nonlinear optics and their applications, and then describe the physical mechanisms, the fabrication, and the characteristics of nonlinear glass materials and several examples of their application.

4.1 Fundamentals of nonlinear optics and applications

4.1.1 General description of nonlinear polarization

Although we are mainly interested in the nonlinear optical effects of glass materials, it will be useful to consider first an elementary definition for nonlinear polarization. For a system of elementary charges e located at the end points of a set of vectors **r** drawn from a common origin, the dipole moment is defined as [3]

$$\mathbf{p} = e\mathbf{r} \qquad (4.1)$$

In a free atom, the center of gravity of the electron distribution coincides with the nucleus and as a whole, the dipole moment in the absence of an external field vanishes. When an electric field is applied, the field tends to draw the center of gravity of the electrons away from the nucleus, resulting in a dipole moment. On the other hand, the attractive forces between the electrons and the nucleus tend to preserve a vanishing dipole moment in the atom. Consequently, the electrons are displaced x slightly from their usual positions and an equilibrium situation is reached. This distortion of the electronic cloud is called the electronic polarization and it is represented schematically in Fig. 4.1.

The medium which concerns us in optics can be thought of as a collection of charged particles: electrons and ion cores. The macroscopic polarization **P** induced in the medium is thought to be a collection of induced electric-dipole moments and if there are N electric dipoles per unit volume, then

$$\mathbf{P} = \sum \mathbf{p} = N\mathbf{p} = \sum \alpha \langle E_{\text{local}} \rangle = \chi \mathbf{E} \qquad (4.2)$$

where α is called the electronic polarizability of electron, χ is the macroscopic susceptibility and **E** is the applied electric field vector [3]. If we consider a

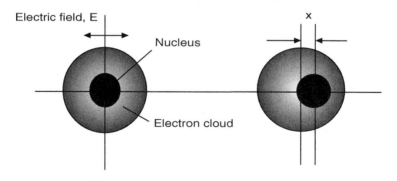

Fig. 4.1.   Schematic illustration of the electronic polarization of an atom (ion) relative to the nucleus under influence of an external field E.

one-dimensional system for simplicity, the effect of the field on a dielectric medium can be expressed in terms of macroscopic susceptibilities χ and applied field E.

4.1.2 Optical nonlinearity of the medium

The motion of the charged particles in a dielectric medium can only be considered to be linear with the applied field if the displacement x is small: the induced electric polarization P varies linearly with the electric field E. For large displacements, however, the restoring force includes additional anharmonic terms and the response is no longer perfectly linear in χ. Then the polarization P may be described by expanding the polarization as a power series of the field E, and the susceptibility $\chi(E)$ of a material is divided into different terms [4–6]

$$P = \chi^{(1)} E + \chi^{(2)} E^2 + \chi^{(3)} E^3 + \cdots$$
$$= \chi(E)E \tag{4.3}$$

$$\chi(E) = \chi^{(1)} + \chi^{(2)} E + \chi^{(3)} E^2 + \cdots. \tag{4.4}$$

Here, $\chi^{(1)}$ denotes the usual linear susceptibility and $\chi^{(2)}$, $\chi^{(3)}$, ... are referred to as nonlinear susceptibilities of the medium.

A light wave consists of electric and magnetic fields which vary sinusoidally at optical frequency. The effect of the optical magnetic field is, however, much weaker and is usually eliminated. For example, if we consider the effect of the alternating electric field of light, $E(t) = E_0 \exp(i\omega t)$, where ω is the optical frequency and E_0 is the field at maximum amplitude, the second terms of Eq. (4.3) correspond to the second-order effects, giving an oscillating polarization that varies with 2ω, resulting in an effect known as second harmonic generation (SHG) [6]. Similarly, the third-order term will give rise to frequency tripled light which is called third harmonic generation (THG).

In a material with inversion symmetry, opposite directions are completely equivalent and the polarization must change sign when the optical electric field is reversed. This condition can be satisfied only if the second-order nonlinear susceptibility $\chi^{(2)}$ is zero. In an isotropic medium such as glass, the nonlinearity is therefore mainly caused by the third-order term of Eq. (4.3) [4, 5] which we will now be considered.

According to the theory of electric and magnetic susceptibilities, the macroscopic quantity χ and, from atomic theory, the dielectric constant are connected by the following relation for a one-dimensional system [3]

$$P = \{(\varepsilon - 1)/4\pi\}E$$

$$= \chi E \tag{4.5}$$

$$\varepsilon = (1 + 4\pi\chi) \tag{4.6}$$

where ε is the static dielectric constant. Now introducing a complex index of refraction, $\mathbf{n} = n - ik$ and a complex dielectric constant, $\varepsilon = \varepsilon' - i\varepsilon''$, whose real part n is the ordinary index and imaginary part k is the extinction coefficient, the relation between them is given by

$$\varepsilon = \varepsilon' - i\varepsilon'' = n^2 = (n - ik)^2. \tag{4.7}$$

When k is small, the index of refraction is given by

$$n^2 = \varepsilon = (1 + 4\pi\chi) = 1 + 4\pi(\chi^{(1)} + \chi^{(3)}E^2). \tag{4.8}$$

The field-dependent part is also frequently expressed in terms of the time-averaged optical electric-field amplitude E as $\langle E^2 \rangle$. The total local index is therefore given by [4, 5, 7, 8]

$$n = n_0 + n_2 \langle E^2 \rangle$$

$$= n_0 + n_2' |E^2|. \tag{4.9}$$

If $|E^2|$ corresponds to $|E_0^2|$, then the two nonlinear coefficients are related by $n_2' = (1/2)n_2$. On the other hand, the phenomenologically observed intensity-dependent refractive index is defined by

$$n = n_0 + n_2'' I \tag{4.10}$$

where I is the average beam intensity in units of W/cm^2, n_0 is the linear refractive index of the material, and the coefficient of proportionality n_2'' is called the nonlinear refractive index. Conversion between the two systems is made with [8]

$$n_2 \text{ (esu)} = (cn_0/40\pi)(n_2'')(m^2/W) \tag{4.11}$$

where c is the velocity of light in vacuum. The variable n_2 is also related to $\chi^{(3)}$ by [4, 8]

$$n_2 \text{ (esu)} = (12\pi/n_0)\chi^{(3)} \text{ (esu)}. \tag{4.12}$$

The dominant nonlinearity in this case is at a frequency well below the band gap. This effect is often referred to as 'non-resonant' in the sense that the excitation does not coincide with an absorption resonance.

If a complex nonlinear susceptibility $\chi^{(3)} = \mathrm{Re}\chi^{(3)} - i\,\mathrm{Im}\chi^{(3)}$ is introduced, the imaginary part of $\chi^{(3)}$ contributes to a change in the absorption coefficient, so that, the absorption coefficient α is a function of light intensity and is described by [9]

$$a = a_0 + \Delta a$$
$$= a_0 + \beta I \tag{4.13}$$

where β is a nonlinear absorption coefficient. From the definition of Eq. (4.7), the absorption coefficient is given by

$$a(\omega) = (4\pi\omega/cn)\,\mathrm{Im}\,\chi. \tag{4.14}$$

Thus, β is related to the imaginary part of the third-order nonlinear suscept-ibility, $\mathrm{Im}\chi^{(3)}$, by

$$\beta = (32\pi^2\omega/c^2 n_0^2)\,\mathrm{Im}\,\chi^{(3)}. \tag{4.15}$$

In this case, excitation above the absorption edge generates predominantly real excited-state populations and the nonlinearities are 'resonant'. The optically-induced changes in the refractive index can be calculated from the optically-induced changes in absorption through the Kramers–Kronig transformation

$$\Delta n(\omega) = (c/\pi)PV \int_0^\infty [\Delta a(\omega')/(\omega'^2 - \omega^2)]\,d\omega' \tag{4.16}$$

where, $\Delta n(\omega)$ and $\Delta a(\omega')$ are the change in the refractive index and in the absorption, respectively, and PV denotes the principal value [10, 12].

4.1.3 Measurement of nonlinear optical properties

4.1.3.1 Degenerate four-wave mixing (DFWM) [13, 14]

The basic geometry of four-wave mixing (FWM) is illustrated schematically in Fig. 4.2 (1). The FWM process involves the two strong counterpropagating pump beams I_1 and I_2, and the weak probe beam I_3 injected at an arbitrary angle to the pump wave. The three beams interact in the nonlinear medium to produce a nonlinear polarization P which then generates the phase conjugate beam I_4. The situation in which all four waves are at the same optical frequency ω is referred to as degenerate four-wave mixing (DFWM).

In the limit of low absorption, the $\chi^{(3)}$ value of the sample is determined according to the following simple equation for the intensity ratio of the phase conjugate beam I_4 to the probe beam I_3, [15, 16]

$$R = I_4/I_3$$
$$= (1024\pi^4\omega^2 I_1 I_2/n^4 c^4)\{\chi^{(3)}\}^2[\{1 - \exp(-\alpha L)\}/\alpha]^2 \exp(-\alpha L). \tag{4.17}$$

Here, α is the absorption coefficient, I_1 and I_2 are the intensity of the forward and backward pump beams respectively, n is the linear refractive index, c is the light velocity, and L is the thickness of the nonlinear medium.

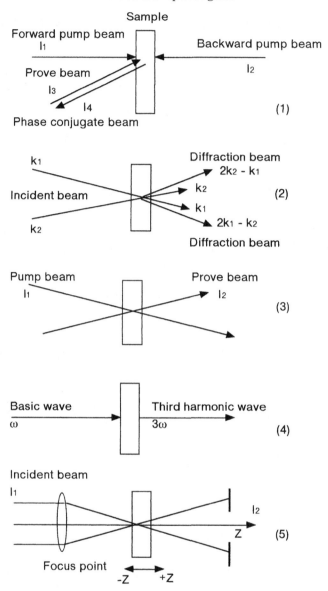

Fig. 4.2. Basic geometry for the measurement of a third order optical nonlinearity. (1) Degenerate four wave mixing; (2) forward degenerate four-wave mixing; (3) pump-prove method; (4) third harmonic generation method; (5) z-scan method.

4.1.3.2 Forward DFWM

A configuration is schematically shown in Fig. 4.2 (2). Two incident beams (wave vectors k_1 and k_2) are focused upon a resonant nonlinear material. For sufficiently small absorption the two beams write the grating by the third-order nonlinear optical effect, which will simultaneously self-diffract and generate two parametrically mixed waves ($2k_2 - k_1$ and $2k_1 - k_2$). Taking into account the absorption α, the diffraction efficiency η, which is defined as the intensity ratio of the diffraction beam to the incident beam I_0, is written in the following equation as [17]

$$\eta = (2^8 \pi^4 \omega^2 / n^4 c^4)\{(1 - T)^2 T / \alpha^2\}|\chi^{(3)}|^2 I_0^2 \tag{4.18}$$

where α and T are the absorption coefficient and the transmissivity, respectively and n and c are the linear refractive index and the velocity of light respectively.

4.1.3.3 Pump-probe technique [18]

A configuration is shown in Fig. 4.2 (3). In the usual experiment, a single frequency pump beam I_1 is tuned above the bandgap of the material and a broad-band probe beam I_2 monitors the pump-induced changes in the transmission of the sample. Then, using Eq. (4.16), the nonlinear susceptibility can be calculated. In this technique, the probe beam is weaker than that of the pump beams, which minimizes the probe-induced effect. In order to discriminate against scattered pump beams, the two beams are perpendicularly polarized and an analyzer which only transmits the probe beams is placed in front of the detector. The change in transmission of the sample induced by the pump beam can be measured as a function of pump fluence. The time dependence of the absorption recovery in the sample is also measured as the delay time between the pump and the probe beams.

4.1.3.4 Maker fringe method (THG method) [19, 20]

Figure 4.2 (4) shows an experimental arrangement for measurement of the relative third-harmonic generation (THG) susceptibility of a sample. The THG intensity $I_{3\omega}$ from the sample can be theoretically expressed as

$$I_{3\omega} \sim I_\omega^3 \{\chi^{(3)}\}^2 L^2 \sin^2(\Delta kL/2)/(\Delta kL/2)^2 \tag{4.19}$$

$$\Delta k = \pi / L_c$$

$$= 6\pi(n_{3\omega} - n_\omega)/\lambda \tag{4.20}$$

where ω is the pump frequency, n is the refractive index of the sample, and L is the effective sample thickness. I_ω and $I_{3\omega}$ are the pump and third-harmonic wave intensity, respectively, $L_c = \lambda/6\Delta n$ is the THG coherence length of the sample, λ is the pump wavelength and $\Delta n = (n_{3\omega} - n_\omega)$ is the refractive index difference between the pump and third harmonic waves. In the experiment, the $\chi^{(3)}$ value of the sample can be determined by comparing the THG peak intensity with the reference sample, assuming that the pump lasers have equal intensities [21, 22].

$$\chi^{(3)} = \chi_r^{(3)} (I_{3\omega}^{1/2}/L_c)/(I_{3\omega,r}^{1/2}/L_{c,r}) \tag{4.21}$$

for $L \gg L_c$, and

$$\chi^{(3)} = (2/\pi)\chi_r^{(3)} (I_{3\omega}^{1/2}/L)/(I_{3\omega,r}^{1/2}/L_{c,r}) \tag{4.22}$$

for $L \ll L_c$. Here, the subindex r refers to the reference sample.

4.1.3.5 Z-scan method [23, 24]

An experimental apparatus is shown in Fig. 4.2 (5). The transmittance of a nonlinear medium is measured through a finite aperture placed in the far field as a function of the sample position (z) with respect to the focal plane. As the sample with negative n_2 and with a thickness much less than the beam depth of focus is placed well in front of the focus ($-z$ in the figure), the increased irradiance leads to a negative lensing effect that tends to collimate the beam, thus increasing the aperture transmittance. With the sample on the $+z$ side of the focus, the negative lens effects tend to augment diffraction, and the transmittance is reduced. A positive nonlinearity ($n_2 > 0$) results in the opposite effect, i.e. lowered transmittance for the sample at negative z and enhanced transmittance at positive z.

A quantity ΔT_{p-v}, defined as the difference between the normalized peak (maximum) T_p and valley (minimum) transmittances T_v, is easily measurable and is given as a function of the average phase distortion $\Delta\Phi_0$ as

$$\Delta T_{p-v} \simeq p|\Delta\Phi_0|, \tag{4.23}$$

for $|\Delta\Phi_0| < \pi$, where $p = 0.406 (1 - S)^{0.25}$, S is the aperture transmittance and $\Delta\Phi_0$ is defined as

$$\Delta\Phi_0 = (2\pi/\lambda)\Delta n_0 \{(1 - e^{-\alpha L})/\alpha\} \tag{4.24}$$

where L is the sample length and Δn_0 is the on-axis index change at the focus ($z = 0$). The nonlinear index n_2 can be calculated using these relationships.

4.1.4 Applications of optical nonlinear materials

One of the applications based on the third-order optical nonlinear effect is the phase conjugate mirror [4, 26–29]. As shown in Fig. 4.2, the phase conjugate beam exactly retraces its original path for the reflection, which will be able to cancel out any phase aberrations within the laser medium.

On the other hand, if we express the intensity-dependent refractive index as

$$n = n_0 + \Delta n(I) \tag{4.25}$$

then, a considerable nonlinear phase change $\Delta\Phi$, given by the following equation, may be achieved,

$$\Delta\Phi(I) = (2\pi/\lambda)\Delta n(I)L \tag{4.26}$$

where L is the optical path length, λ is the signal wavelength [25]. As an example, for $\lambda = 1$ μm and $\Delta n = 10^{-4}$, a phase change of 2π is obtained after 1 cm propagation.

A typical example of devices that are derived from these phase-changes is a Mach–Zehnder interferometer pulse selector [25]. As shown in Fig. 4.3(a), the nonlinear interferometer consists of Y-junctions and single channel input and output waveguides. If the two split channels are identical, the signals arrive at the second Y in the same phase and sum to produce an output signal that is equal to the input signal, even if the intensity-dependent effects alter the phase in both channels. If asymmetry is introduced as a difference in channel length, then the nonlinear phase changes in the two channels will not be equal and the output power will be nonlinearly dependent on the input power. Hence the input signal is selected by light intensity. If the directional coupler, by which the power is coupled between two closely spaced waveguides as shown in Fig. 4.3 (b), is fabricated in a nonlinear material, then the coupling condition will be intensity-dependent and pulse selection is also possible [25, 30].

Another example is the Fabry–Pérot (FP) interferometer which is shown in Fig. 4.3 (c) [31–33]. The FP transmission curve T versus phase change $\Delta\Phi$ is described by

$$T = T_0/\{1 + F\sin^2(\Delta\Phi/2)\} \tag{4.27}$$

where T_0 and F are constants. If the FP interferometer is filled with a nonlinear medium, then the transmission of a monochromatic beam through the interferometer depends on the beam intensity because of the optical-field-induced refractive index.

The optical Kerr effect can be used in optical switching devices, as shown in Fig. 4.3 (d). The transmission of the weak signal beam passing through the nonlinear medium is normally blocked by the crossed analyzer. In the presence of an intense polarized beam (gate pulse), however, the medium becomes birefringent, and the signal beam experiences a polarization change in traversing the medium and is no longer completely blocked by the analyzer. Therefore, by using a picosecond pulse as a gate pulse, a nonlinear medium that has a picosecond response time can act as a fast optical shutter for a signal beam which has a picosecond on-off time.

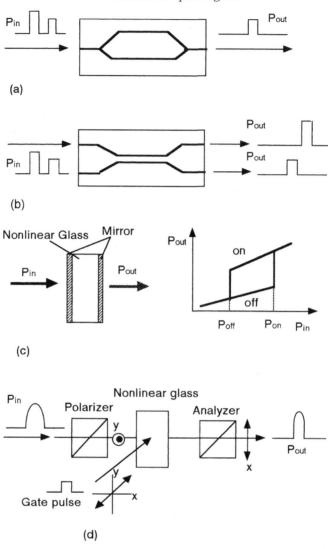

Fig. 4.3. Standard nonlinear optical devices and their response to optical power, (a) Mach–Zehnder interferometer – pulse selector. (b) Nonlinear directional coupler. (c) Fabry–Pérot interferometer – bistable switching and (d) Optical Kerr shutter.

4.2 Nonresonant nonlinear glasses

The third-order nonlinearity is present in all glass materials, and the dominant nonlinearity is produced by the excitation in the transparent frequency region well below the band gap. This effect is intrinsic and often referred to as 'non-resonant'.

Various types of glass are most commonly used as the optical materials, and are not nonlinear for most applications. At high intensities, however, the optical nonlinear effect reduces the focal intensity or induces self-focusing which often leads to more serious local damage to the optical materials. Early studies of optically induced changes of refractive index were aimed at finding glasses with small nonlinearities in order to avoid self-focusing in high-power optical components and laser crystals [34]. Several oxyfluoride glasses and fluoride glasses have been extensively studied for possible use in high-power laser applications because of their small optical nonlinearities.

At the opposite extreme, other recent investigations have been directed towards the largest possible nonlinearities, for all-optical signal-processing and switching devices as mentioned in Section 4.1.4. Optical glasses with high refractive indices and high dispersions are good candidate materials for these applications. In addition, different glass systems are now under investigation with the aim of increasing their optical nonlinearities by introducing a variety of modifiers into the glass network.

4.2.1 Nonlinearity of dielectric materials

The total nonlinear refractive index of nonresonant glass materials includes both electronic and nuclear contributions. The fast component of n_2 is believed to arise from electronic effects. In the presence of a strong electromagnetic wave, the atom is polarized. For most transparent dielectrics, the third-order nonlinearity results from the anharmonic terms of the polarization of bound electrons.

A useful model for the intrinsic nonlinearity of the glasses can be developed from the equation of motion of, for example, a one-dimensional anharmonic oscillator. If every electron of charge e and mass m is supposed to be bound in its equilibrium position by an elastic force, the equation of motion of each oscillator is given by the following equation [2, 35],

$$m(\mathrm{d}^2x/\mathrm{d}t^2) + m\gamma(\mathrm{d}x/\mathrm{d}t) + m(\omega_0^2 x + \mu x^3) = -eE \qquad (4.28)$$

where x is the displacement from the mean position, $\omega_0 = (f/m)^{1/2}$ is the resonance frequency, f is the restoring force constant, γ is the damping constant, μ is the anharmonic force constant, E is an applied electric field and eE is the electrostatic force. The restoring force includes a term that is third-order in displacement. Ignoring this third-order displacement, and taking the form of an alternating field $E = E_0 \exp(i\omega t)$ as the applied electric field, the movement of the electrons can be defined by the equation of harmonic motion. A solution of this equation is [3, 36]

$$x(t) = (-eE/m)/\{(\omega_0^2 - \omega^2) + i\gamma\omega\}. \tag{4.29}$$

The polarization induced in the medium and the related parameters can be immediately obtained from Eqs (4.1), (4.2), (4.5) and (4.6):

$$P = -Nex$$

$$= (eE/m)/\{(\omega_0^2 - \omega^2) + i\gamma\omega\}$$

$$= \chi E \tag{4.30}$$

$$\varepsilon = 1 + 4\pi\chi^{(1)}$$

$$= 1 + (4\pi Ne^2/m)/\{(\omega_0^2 - \omega^2) + i\gamma\omega\}. \tag{4.31}$$

Now, separating ε into its real and imaginary components, and also assuming $\omega \ll \omega_0$ and $k \ll 1$, then

$$\varepsilon = n^2$$

$$= 1 + (4\pi Ne^2/m)/(\omega_0^2 - \omega^2) \tag{4.32}$$

$$\chi^{(1)}(\omega) = (Ne^2/m)/(\omega_0^2 - \omega^2). \tag{4.33}$$

On the other hand, if we assume $\omega \ll \omega_0$, Eq. (4.28) may also be solved for a third-order nonlinear susceptibility when one nonlinear term dominates. Then, a relationship analogous to Miller's rule for second-order nonlinear susceptibilities [37]

$$\chi^{(3)} = (m\mu/N^3 e^4)[\chi^{(1)}(\omega)]^4 \tag{4.34}$$

is obtained for centrosymmetric materials such as glasses [2].

A more detailed calculation of the optical nonlinearity in terms of the parameters of linear refractive index and dispersion can be carried out in order to develop low n_2 or high n_2 materials. The measurement of the nonlinear refractive index coefficient n_2 involves fairly complicated experimental arrangements and is a time-consuming task. Ignoring the second term of Eq. (4.28) and taking the local field effects into consideration, C. C. Wang [38, 39] developed an alternative relationship to Eq. (4.34) for ionic crystals, at low frequency ($\omega \sim 0$):

$$\chi^{(3)} \sim g[\chi^{(1)}]^2 f^3 / N_{\text{eff}} \hbar \omega_0 \tag{4.35}$$

where $\chi^{(1)}$ is the linear susceptibility given by Eq. (4.33), f is a local field correction factor which is usually taken to be the Lorentz local field $f = (n^2 + 2)/3$, N_{eff} is the oscillator strength or the number of effective oscillators and ω_0 is the mean absorption frequency. The factor g is the 'anharmonicity' parameter and is a dimensionless quantity involving the properties of the ground state as well as the excited states of the system; in most cases this factor is not directly amenable to measurement. Using the known values of the $\chi^{(1)}$, N_{eff} and ω_0, he calculated $\chi^{(3)}$ from Eq. (4.35) and compared it with the experimental values of $\chi^{(3)}$ for several gases. This empirical relation held within the available experimental accuracy for gases at low pressures, showing that $g = 1.2$ gave a better fit to the data.

4.2.2 *BGO model [35] for nonresonant optical nonlinearities*

N. L. Boling, A. J. Glass and A. Owyoung (BGO) showed that the relation (4.35) can be generalized to a wide variety of solids, including some oxide glasses. The semi-empirical formula that they proposed has been a useful guide in the search for low and high nonlinear refractive index glasses. Applying the familiar Claussius Mosotti equation in the case that one polarizable constituent of the medium dominates, they wrote the nonlinear refractive index in the low frequency limit ($\omega \ll \omega_0$) as

$$n_2 = (gs)(n^2 + 2)^2(n^2 - 1)^2 / 48\pi n\hbar\omega_0(Ns) \tag{4.36}$$

where s is the effective oscillator strength. In the classical model, the quantity g is related to the anharmonic coefficient μ in Eq. (4.30) [35] as

$$g = \mu s\hbar / m\omega_0^3. \tag{4.37}$$

The refractive index can be written as

$$\{(n^2 - 1)/4\pi\}\{3/(n^2 + 2)\} = (Ns)e^2/m(\omega_0^2 - \omega^2). \tag{4.38}$$

Thus, in principle, the parameter Ns and ω_0 can be obtained from any two values of $n(\omega)$. A large group of optical glasses were examined on this basis and a value for gs of about 3 was found to give good agreement for all materials examined.

N. L. Boling *et al.* [35] also proposed specifically an expression connecting the nonlinear refractive index n_2 with the optical parameters such as n_d and the Abbe number v_d as follows

$$n_2(10^{-13} \text{ esu}) = 68(n_d^2 + 2)^2(n_d - 1)/v_d[1.52 + \{(n_d^2 + 2)(n_d + 1)v_d\}/6n_d]^{1/2}. \tag{4.39}$$

This expression has successfully related n_2 at 1.06 μm with n_d and v_d for weakly dispersive optical materials [34, 35, 40, 41].

4.2.3 *Lines' model for nonresonant optical nonlinearities*

The theories of nonlinear response in glasses that have been discussed above are semi-empirical and only concern the anharmonic electronic response associated with the individual ion themselves. M. E. Lines [42, 43] introduced the concept of a bond-orbital theory which led to a semiquantitative representation of the nonlinear response in optically transparent materials as functions of such readily available measures as formal valency, bond length, ionic radii, etc. Bond orbital theory, in its simplest form, describes the long-wavelength electronic response as the perturbation of local bonding orbitals by an applied electric field. Finally, the average nonlinear refractive index $n_2\,(av)$ can be written in terms of measurable physical quantities as the very simple proportionality:

$$n_2(av) = 25 f_{\rm L}^3 (n_0^2 - 1)d^2 E_{\rm s}^6 / n_0 [E_{\rm s}^2 - (\hbar\omega)^2]^4\,(10^{-13}\ {\rm cm}^3/{\rm erg}) \qquad (4.40)$$

where d is the bond length, $f_{\rm L}$ is the local-field factor which is usually assumed to be the 'Lorentz local field factor' $(n_0^2 + 2)/3$, $E_{\rm s}$ in eV is the common single oscillator Sellmeier gap. This is observable from the long-wavelength electronic frequency dependence of the linear response in the form

$$(n_0^2 - 1) \sim \{E_{\rm s}^2 - (\hbar\omega)^2\}^{-1}. \qquad (4.41)$$

A direct comparison of his theory with experiment for relative n_2 value over 11 halides showed an rms accuracy of 9%.

The nonlinear optical properties of various glasses have been discussed on the basis of Lines' model. The measured $\chi^{(3)}$ values of TeO_2 glasses reasonably agree with the calculated $\chi^{(3)}$ values derived by Eq. (4.40) [44]. However, this model could not be applied to Sb_2O_3–B_2O_3 glasses [45] and BaO–TiO_2–B_2O_3 glasses [46], because it is difficult experimentally to evaluate the proper bond length (d) for these glasses. By using Wemple's single oscillator approximation [47, 48] between the dispersion energy ($E_{\rm d}$) and the bond length, $E_{\rm d} \sim d^2$, a modified Lines' equation for $\chi^{(3)}$ was proposed as [45, 46]

$$\chi^{(3)} \sim (n_\omega^2 + 2)^3 (n_\omega^2 - 1)E_{\rm d}/E_0^2 \qquad (4.42)$$

where n_ω is refractive index, $E_{\rm d}$ is a special parameter called the 'dispersion energy' and E_0 is the energy of the effective dispersion oscillator, typically corresponding to the single oscillator Sellmeier gap $E_{\rm s}$. These three optical parameters n_ω, $E_{\rm d}$ and E_0 can be determined experimentally from refractive index dispersion data using [47]

$$(n(\omega)^2 - 1) = E_{\rm d}E_0 / \{E_0^2 - (\hbar\omega)^2\}. \qquad (4.43)$$

4.2.4 Nuclear contributions

The magnitude of the nuclear contribution to the optical nonlinearities was estimated for a number of glasses [49, 50]. From these results, it was concluded that the nuclear contribution is $15 \sim 20\%$ of the total response for fused silica [49] and $12 \sim 13\%$ for sulfide glasses and heavy-metal oxide glasses [50].

4.2.5 Nonlinear optical properties of nonresonant nonlinear glasses

4.2.5.1 Nonlinear refractive index

The semi-empirical expressions for n_2 derived by N. L. Boling *et al.* [35] have been very useful as a guide for developing relatively high n_2 materials. The conclusion that can be drawn from their expressions is that high index and high dispersion materials should possess large nonlinearities. In commercial optical glasses, the reported measurements of n_2 for the Pb^{2+}-containing glasses (SF series) are higher than those of other optical glasses, as summarized in Table 4.1. The expression with $gs = 3$ has been quite successful in predicting the nonlinear refractive indices of materials with small nonlinearities [34, 35]. In Fig. 4.4, the measured values of n_2 reported for a number of optical glasses and special glasses for laser systems are plotted against computed values when $gs = 3$ and show good agreement.

R. Adair *et al.* also measured the nonlinear refractive indices of a large number (more than 90) of optical crystals [63, 64] and several tens of glasses [54] using a form of DFWM and they compared the results with the predictions of the BGO formula. From measured values of n_2 obtained on a variety of oxide glasses, a value of $gs = 3$ yielded the best fit, as in the case of the BGO results for low dispersive glasses. However, a value of $gs = 2$ gave a better fit to their data on optical crystals than $gs = 3$ [63]. More recently, Kang *et al.* [65] showed that their experimental results were fitted best with the value $gs = 1$ and concluded that the value of gs is material-dependent in principle and changes as the magnitude of the nonlinearity changes.

Systematic studies of the relationship of optical nonlinearities and glass compositions have also been conducted in many glass systems. In oxide glasses, for example, high linear and nonlinear refractive indices are found in systems containing large, polarizable cations such as Ti, Bi, Tl, Pb, Nb or Te. Table 4.2 summarizes representative n_2 values for a number of different glasses measured using several different techniques.

The relationship between the linear and the nonlinear refractive indices described above is of course cation-dependent. E. M. Vogel *et al.* [1, 79] have summarized some of the results in their reviews. For instance, ions with an empty or unfilled d shell such as Ti and Nb contribute most strongly to the linear and nonlinear polarizabilities. However, in glass which contains TiO_2 and Nb_2O_3, the linear refractive index is mainly determined by the total concentration of Ti and Nb, whereas the nonlinear index coefficient is much larger for glasses that contain TiO_2 than for those containing Nb_2O_3 [66]. K. Terashima *et al.* [46] also measured the third-order nonlinear optical suscept-ibilities of TiO_2-containing glasses and suggested that the linear optical proper-

Table 4.1. *Nonlinear refractive indices and nonlinear susceptibilities of optical glasses and special glasses for laser systems*

Glass type	Refractive index n	Nonlinear refractive index n_2 (10^{-13} esu)	Nonlinear refractive index n_2'' (10^{-16} cm^2/W)	Nonlinear susceptibility $|\chi^{(3)}|$ (10^{-13} esu)
Optical glass				
Fused silica	1.456(d) [34], 1.458(D) [51] (d) [35], 1.452(d) [53]	0.99 [34], 0.95* [35, 41, 51], 0.85* [54]	2.73* [51]	0.28 [44, 59, 60]
FK–6	1.446(d) [34]	1.13 [34]		
BK–7 (borosilicate)	1.517(d) [34] (D) [51]	1.46 [34], 1.24* [35, 51], 1.30* [54]	3.43* [51]	
SF–7 (PbO-silicate)	1.640(d) [34]	5.83 [34]		
LaK–3	1.694(d) [34]	2.57 [34]		
LaSF–7	1.914(d) [34], 1.85(d) [58]	6.20 [34]		0.45 [58]
FK–5 (fluorosilicate)	1.487(D) [51] (d) [35]	1.07* [35, 51]	3.01* [51]	
FK–51 (fluorosilicate)	1.487(D) [51]	0.69* [35, 51]	1.94* [51]	
SF–59 (PbO-silicate)	1.953(d) [53, 56], 1.91* [55, 57], 1.97 [61], 1.917 [53]		68* [55], 32 [57], 67.6 [62], 90 [61]	0.78# [53] 1.11# [53], 0.75 [56, 57], 0.66 [58]
SF–57 (PbO-silicate)	1.8467(d) [53], 1.81* [55]		41* [55]	0.51# [53], 0.225 [58]
SF–6 (PbO-silicate)	1.8052(d) [53, 56], 1.77* [57]	8.0* [54]	22* [57]	0.45# [53], 0.44 [56, 57], 0.165 [58]
FD–60 (PbO-silicate)	1.8052(d) [53], 1.77* [55]		20* [55]	0.42# 53
LaSF–30	1.8032(d) [53]			0.12# [53]
SF–58	1.917(d) [56], 1.88* [55, 57]		49* [55], 23 [57]	0.52 [56, 57], 0.495 [58]
SF–56	1.75* [55]		26* [55]	
FDS–9	1.81* [55]		12* [55]	
FD6	1.77* [55]		31* [55]	
FDS–90	1.81* [55]		22* [55]	
8463 (Corning)	1.94 [57], 1.97(D) [58]		42* [57]	1.00 [57], 0.825 [58]
Special Oxide glass				
FR–5 (Tb$_2$O$_3$-aluminosilicate)	1.686(D) [51], 1.678* [52]	2.1 [35, 51], 1.93* [54]	5.2* [51]	
FR–4 (Ce$_2$O$_3$-phosphate)	1.556* [52]	1.95* [52]		

(d): d line; (D): D line; *: 1.06 μm; #: 0.65 μm.

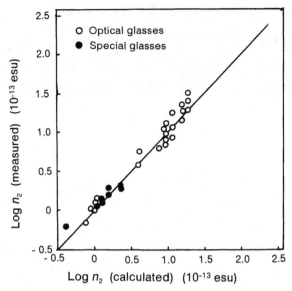

Fig. 4.4.   Plot of measured n_2 vs the value of n_2 calculated by BGO model (data from Tables 3.5 and 4.1).

ties are determined by the formation of the TiO_6 unit, whereas the nonlinearities are determined by the amount of net Ti^{4+} ions.

In general, the nonresonant type optical nonlinearity is derived from the hyperpolarizabilities of the glass constituents, such as the heavy metal cations (Pb^{2+}), transition-metal ions (Ti^{4+} or Ti^{3+}) and Te^{4+}. The high nonlinearity of Pb^{2+}-containing glasses is attributed to the unusually large hyperpolarizabilities of the nonbonding lone electron pairs. Furthermore, the many filled inner-electronic shells with lead ions screen the outer electrons effectively from the nucleus, thus promoting large charge displacements to occur under the influence of an optical field. The anharmonic effects are thought to arise from such large displacements [75]. The high nonlinear refractive indices of TiO_2 glass and TeO_2 glass are attributed to the large number of low-lying, empty (Ti^{4+}) and sparsely occupied (Ti^{3+}) 3d electronic states [66], or to the empty 5d orbitals of the Te ions [69], respectively.

It was also suggested that the positive nonresonant-type optical nonlinearity derived from the high polarizabilities of the glass constituents, such as Ti^{4+} would be cancelled by the negative refractive index resulting from the contributions of the resonant 3d electronic transition of Ti^{3+} ions (540 nm and 660 nm), for instance, in TiO_2-containing glasses [75].

As shown in Table 4.3, chalcogenide glasses are found to have the largest nonresonant third-order nonlinearities reported to date [80]. For the nonlinear

Table 4.2. *Nonlinear refractive index for various high index oxide glasses*

Glass type	Refractive index	Nonlinear refractive index n_2 (10^{-13} esu)	Nonlinear refractive index n_2'' (10^{-16} cm^2/W)	Nonlinear susceptibility $|\chi^{(3)}|$ (10^{-13} esu)	Nonlinear susceptibility $|\chi^{(3)}_{1111}|$ (10^{-13} esu)	Measurement (μ, μm)	Ref.
K−1261 (79%TeO$_2$)		23*				TWM (1.06)	54
3151 (88%TeO$_2$)		24*				TWM (1.06)	54
RN(40PbO−35BiO$_{1.5}$−25GaO$_{1.5}$)	2.46 (D)		125		4.2	DFWM (1.06)	56
	2.30*		125		4.2	DFWM (1.06)	57
	2.46 (d)				4.2*, 380(0.6),		65
NbO$_{2.5}$−TiO$_2$−Na$_2$O−SiO$_2$	1.95 (0.633)		40.3			DFWM (1.06)	66
QR(PbO-silicate)	2.02*		43		1.1	DFWM (1.06)	57
QR(PbO-silicate)	2.06 (D)				0.86	KE (0.633)	58
VIR−3 (Germanate)	1.84*		22		0.48	DFWM (1.06)	57
QS(PbO−germanate)	1.94*		34		0.80	DFWM (1.06)	57
E1(TiO$_2$−Nb$_2$O$_5$−borosilicate)	1.96 (0.633)		40 (0.633)				67
QZ(Tl$_2$O−CdO−PbO−Bi$_2$O$_3$−Ga$_2$O$_3$)	2.27 (D)				3.60	KE (0.633)	58
QY(Tl$_2$O−CdO−PbO−Ga$_2$O$_3$)	2.31 (D)				4.20	KE (0.633)	58
BZA(BaO−ZnO−TeO$_2$)	2.17 (D)				1.65	KE (0.633)	58
BZA	2.17 (D)				1.5	DFWM (1.06)	56
AAF(TiO$_2$−Nb$_2$O$_5$-borosilicate)	1.93 (D)				2.685	KE (0.633)	58
ACU(PbO−TiO$_2$-silicate)	1.86 (D)				0.495	KE (0.633)	58
ACU					0.435	DFWM (1.06)	56
PbO−TiO$_2$−TeO$_2$	~2.27 (0.589)			~37.0		THG (1.9)	59
Nb$_2$O$_3$−TiO$_2$ −TeO$_2$	~2.33 (0.598)			~9.6		THG (1.9)	59
Li$_2$O−TiO$_2$−TeO$_2$	2.2 (0.598)			8.00		THG (1.9)	68
PbO−GaO$_{1.5}$	2.26 (0.635)			7.7		THG (1.907)	60
TiO$_2$−PbO−GaO$_{1.5}$	2.301 (0.635)			8.5		THG (1.907)	60
Nb$_2$O$_5$−PbO−GaO$_{1.5}$	2.280 (0.635)			7.5		THG (1.907)	60
WO$_3$−PbO−GaO$_{1.5}$	2.258 (0.635)			6.7		THG (1.907)	60
TeO$_2$	2.184 (0.633)			14.1		THG (1.9)	69
DZ(Tl$_2$O−Bi$_2$O$_3$−Ga$_2$O$_3$)	2.47 (d)				5.6*, 520 (0.6)	Z	65
Pb9(PbO−TeO$_2$−SiO$_2$)	2.27 (d)				4.0*, 57 (0.6)	Z	65

Material	n			Method	Ref
PbBi(PbO–Bi$_2$O$_3$–borosilicate)	2.34 (d)		3.1*, 290 (0.6)	Z	65
Sb$_2$O$_3$–B$_2$O$_3$	~1.96 (1.9)		8.4	THG (1.9)	45
Bi$_2$O$_3$–SiO$_2$–TiO$_2$	1.87 (0.656)		58	THG (1.9)	70
Bi$_2$O$_3$–B$_2$O$_3$–SiO$_2$	2.05 (0.656)		93	THG (1.9)	70
TeO$_2$ glass	2.0*	8.1*, 43 (0.532)		Z (1.06, 0.532)	71
Titanium-borophosphate	~1.8	~69		PP	61
Al$_2$O$_3$–TeO$_2$	~2.04	~174		PP	61
Ag$_2$O–B$_2$O$_3$	~1.631 (0.633)		~1.22	THG (1.9)	72, 73
AgCl–Ag$_2$O–B$_2$O$_3$	~1.907 (0.633)		~4.20	THG (1.9)	72
AgBr–Ag$_2$O–B$_2$O$_3$	~2.005 (0.633)		~13.4	THG (1.9)	72
AgI–Ag$_2$O–B$_2$O$_3$	~2.179 (0.633)		~60.2	THG (1.9)	72
Na$_2$O–TiO$_2$–P$_2$O$_5$(35%TiO$_2$)		~65.8 (0.8)		PP	62
La$_2$O$_3$–TiO$_2$–GeO$_2$			3.73	THG (1.9)	74
PbO–TiO$_2$–SiO$_2$		77.2		Z (0.532)	75
BaO–TiO$_2$–B$_2$O$_3$	~1.94 (0.633)		~3.04	THG (1.9)	46
Na$_2$O–TiO$_2$–P$_2$O$_5$	~1.76	48		PP (0.8)	76
Na$_2$O–TiO$_2$–B$_2$O$_3$–P$_2$O$_5$	~1.80	69		PP (0.8)	76
Na$_2$O–TiO$_2$–SiO$_2$	1.75	32		PP (0.8)	76
La$_2$O$_3$–B$_2$O$_3$	1.67 ~ 1.72 (0.633)		0.6 ~ 1.0	THG (1.9)	77
Pr$_2$O$_3$–B$_2$O$_3$	1.67 ~ 1.78 (0.633)		15.1 ~ 18.6	THG (1.9)	77
Nd$_2$O$_3$–B$_2$O$_3$	1.72 ~ 1.78 (0.633)		0.7 ~ 1.5	THG (1.9)	77
Sm$_2$O$_3$–B$_2$O$_3$	1.72		1.2 ~ 1.5	THG (1.9)	77
PbO-Silicate	–		21 ~ 32	PP (0.63)	78

(d): d line, (D): D line; *: 1.06 µm; (λ), wavelength in µm; TWM, three-wave mixing; DFWM, degenerate four-wave mixing; KE, optical Kerr Effect; THG, third harmonic generation; TRI, time resolved interferometry; Z, z scan; PP, pump probe.

Table 4.3. *Nonlinear refractive indices of chalcogenide glasses*

Glass type	Refractive index (λ, μm)	Nonlinear refractive index n_2'' (10^{-16} cm^2/W)	$|\chi^{(3)}|$ (10^{-13} esu)	Nonlinear susceptibility $|\chi^{(3)}_{1111}|$ (10^{-13} esu)	Measurement (λ, μm)	Ref.
As_2S_3	2.48 (1.06)		14.8 (0.635)		DFWM	80
	2.53 (2)		22		THG (1.9)	80
	2.4 (1.06)			17.4	DFWM (1.06)	57
		1400	72		THG (2)	81
			100		Z (1.06)	82
			120		THG (1.9)	70
	2.38	420 (1.3)			KE (1.303)	83
		40 (1.3)				88
GeS_3			10		THG (1.9)	80
$90GeS_2-10Tl_2S$			7		THG (1.9)	80
$90As_2S_3-10I$			3		THG (1.9)	80
Ge–S(NOG–1)	2.05 (1.9)		20		THG (1.9)	81
Ge–As–S(NOG–2)	2.22 (1.9)		37		THG (1.9)	81
Ge–As–S(NOG–3)	2.35 (1.9)		46		THG (1.9)	81
Ge–As–S–Se(NOG–4)	2.36 (1.9)		70		THG (1.9)	81
As–S–Se(NOG–5)	2.55 (2.0)		141		THG (1.9)	81
As–S–Se	~2.47 (10.6)	~175 (1.55) 1.02×10^5			THG (2)	88
$Ag_2As_{39}S_{59}$	2.43 (1.06)			220 (0.6), 7.9 (1.25)	Z (1.06)	82
VA(La_2S_3–Ga_2S_3)	2.50			190 (0.6), 2.2 (1.25)		65
ACN(GeS_2–Ga_2S_3–BaS)	2.22					65

Composition	n (λ)				Method (λ)	Ref
XT(GeS$_2$–Ga$_2$S$_3$)	2.19			300 (0.6), 2.7 (1.25)		65
As$_{40}$S$_{57}$Se$_3$	2.5 (0.656)		140		THG (1.9 ~ 2.2)	70, 85, 87
Na$_2$S–GeS$_2$	2.03 ~ 2.08 (1.9)		~50		THG (1.9)	84
(PbS, Ag$_2$S, Na$_2$S)–La$_2$S$_3$–Ga$_2$S$_3$	2.13 ~ 2.41 (0.633)		11.1 ~ 32.4		THG (1.9)	86
Ge–As–Se–Te	2.50 (10.6)	140 (1.3)				88
Ge–As–S–Se	2.41 (10.6)	35 (1.55)				88
As–Se–Sb–Sn	2.80 (10.6)	150 (1.33)				88
Ge–Sb–Se	2.60 (10.6)	80 (1.3)				88
GeS$_2$–Ga$_2$S$_3$–CsI–Ag$_2$S	1.99 ~ 2.16 (0.633)	165 ~ 750			Z (0.705)	89
As$_2$S$_3$		250			Z (1.00)	90
GeSe$_4$		800			Z (1.06)	90
Ge$_{10}$As$_{10}$Se$_{80}$		1020			Z (1.06)	90

THG, third harmonic generation; DFWM, degenerate four-wave mixing; Z, z scan; KE, optical Kerr effect.

susceptibility of chalcogenide glasses, in which the minimum intrinsic absorption is reported to occur at about $4 \sim 6$ μm compared to about ~ 1 μm in oxide glasses, there is an uncertainty of the resonant contribution to the reported values of $\chi^{(3)}$ even at 2 μm. In four-wave mixing experiments, D. H. Hall *et al.* [57] suggested that a large mixing signal may be due to a significant imaginary part of $\chi^{(3)}$, which is probably caused by the onset of resonant nonlinear processes such as two-photon absorption.

4.2.5.2 Nonlinear response time

The total nonlinear response of the nonresonant glass materials includes both electronic and nuclear contributions [49, 50, 54]. The electronic term is due to the deformation of the electron orbits. The process is a virtual population process as long as the field is applied, so that there is no energy deposition in the material and no participation in a relaxation process. The nonresonant nonlinearity process is expected to have a nearly instantaneous response of several electronic cycles ($\sim$fs) [49, 91].

I. Thomazeau *et al.* [53] measured and resolved the nonlinear response of several inorganic glasses on a femtosecond time-scale using the optical Kerr-effect technique. The results indicate that the responses may be attributed to electronic processes whose characteristic response times are less than 100 fsec. These results also mean that the nuclear nonlinearities are either too fast or too small to be detected using this technique, but the contribution from the nuclear motion is not yet resolved [65]. S. R. Friberg *et al.* [55] also reported DFWM measurements of the nonlinear response of a number of optical glasses at 1.06 μm and obtained the response time of 1 ps for SF59 optical glass.

M. Asobe *et al.* [83] recently demonstrated ultrafast all-optical switching in a Kerr-shutter configuration using a single mode As_2S_3-based glass fiber that was 48 cm long. In this experiment they estimated a switching time of less than 15 ps and confirmed later that the response time was less than 100 fs using the pump-probe technique [92]. Ultrafast responses below 200 fs were also reported recently in Bi_2O_3-containing glasses [70] and chalcogenide glasses [87]. Table 4.4 lists the nonlinear response times of nonresonant glass materials.

4.2.5.3 Applications

Practical applications of nonresonant glass materials in ultrafast all-optical switches have been reported by M. Asobe *et al.* using chalcogenide glass fibers [83, 92–94]. Figure 4.5 shows the switching characteristics of a 1.2-m long

Table 4.4. *Nonlinear response times of nonresonant glass materials*

Glass	λ (nm)	$\chi^{(3)}$ (esu)	n_2'' (cm^2/W)	τ (ps)	Ref.
SF59	650	$2.6 \sim 3.7 \times 10^{-14}$		0.17	53
SF59	1060		7×10^{-15}	1	55
As$_2$S$_3$	1319	1.2×10^{-11}	4.2×10^{-14}	5	83
As$_2$S$_3$	1552		2×10^{-14}	0.1	92
Bi$_2$O$_3$ Glass	810	$5.8 \sim 9.3 \times 10^{-12}$		0.2	70
As$_{20}$S$_{80}$	810			0.2	87

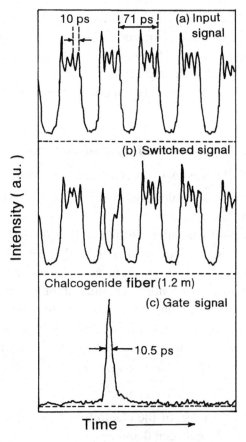

Fig. 4.5. 100 GHz switching characteristics of 1.2 m long chalcogenide fiber with Kerr shutter configuration. (a) input signal, (b) output signal, (c) gate signal. [Reprinted from M. Asobe and T. Kanamori, Proc. International Workshop on Advanced Materials for Multifunctional Waveguides, Chiba (1995) 64, with permission from U. A. Army Research Office.]

As_2S_3 fiber with Kerr-shutter configuration, which demonstrates all-optical 100 GHz switching performance using a laser diode as a gate pulse source.

4.3 Semiconductor-doped glasses

Glasses that have been doped with microcrystallites of semiconductors are serious candidates as resonant nonlinear materials. Their rapid response times of a few tens of picoseconds makes them attractive for nonlinear device applications. One type of these materials that has been available commercially as sharp-cut color filters and manufactured by glass-makers for many years, is one in which a CdS_xSe_{1-x} microcrystallite phase is thermally developed in the glass matrix. The importance of semiconductor-doped glasses for nonlinear optical applications was realized first by R. K. Jain and R. C. Lind in 1983 who reported DFWM in these glasses [14]. Since then, the third-order nonlinear optical properties of semiconductor-doped glasses have been extensively investigated. For the specific case of resonant nonlinearities, the exciton state in the microcrystallite form of wide gap semiconductors such as CuCl and CuBr doped in the glass matrix produces different optical nonlinear effects [95–98].

4.3.1 Fundamentals of semiconductors

4.3.1.1 Band theory of semiconductor [99–102]

Before discussing the actual problems of semiconductor-doped glasses, it may be useful to understand the elementary theory of the semiconductor itself. The band structure of a crystal can often be treated physically by the nearly free electron model as a simple problem of a linear solid. According to the quantum theory, the state of an electron in an atom is characterized by the principal quantum number n, angular momentum quantum number l, magnetic quantum number m and spin quantum number s. To simplify the discussion, it is assumed that an electron is confined uni-directionally to a length L by infinite potential energy barriers. The allowed energy value of the electron is given by

$$E_n = (\hbar^2/2m)(n\pi/L)^2$$
$$= (\hbar^2/2m)k^2 \qquad (4.44)$$

where $k = n\pi/L$, m is the mass of the electron in free space, n is an integral number of half-wavelengths between 0 and L.

The band formation arising from the crystal lattice is analogous with the electronic state in an atom. A crystal is a periodically repeated array of atoms. When two atoms are very far apart each with its electron in the 1s state there is no overlapping between neighbouring bands. The allowed levels consist of a single doubly degenerate level, i.e. each electron has precisely the same energy. As the atoms are brought closer together, their wave functions overlap and form two combinations of wave functions as shown in Fig. 4.6 (a). Now, as six free atoms are brought together, standing waves of electrons passing along a cubic crystal of length L_c should satisfy the Schrödinger wave equation and the degenerate level will be split into six by the interaction between the atoms, as

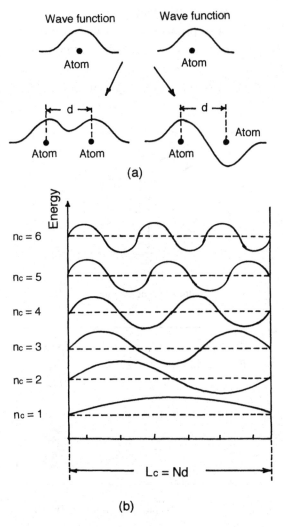

Fig. 4.6. (a) Combination of two wave functions, (b) Energy levels and wave functions of six electrons in the crystal of length L_c.

shown in Fig. 4.6 (b) for the case of six unit cells and $n = 1$. The allowed energy of the electron in the crystal is therefore given by

$$E_c = (\hbar^2/2m)(n_c\pi/L_c)^2$$
$$= (\hbar^2/2m)k_c^2 \qquad\qquad (4.45)$$

where $k_c = (n_c\pi/L_c)$ and $n_c = 1, 2, \ldots, N$, $L_c = Nd$ as d is the inter-atomic distance in a one-dimensional case and N is the number of atoms in the crystal. We therefore have a band of N closely spaced levels instead of a single degenerate level, as shown in Fig. 4.7. A wave function with $k_c = \pi/L_c$ must represent a standing wave with minimum E_c.

The width of the band is proportional to the strength of the interaction or overlap between neighboring atoms. As the atoms approach, the splitting between the levels will be increased as shown in Fig. 4.7 (a). For the deep-lying levels the perturbation will be small compared with the attractive force of the

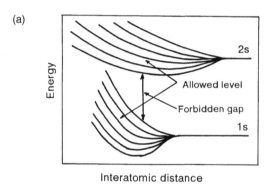

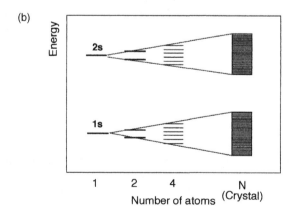

Fig. 4.7.   (a) Energy bands in crystalline solid as a function of lattice spacing, (b) Energy levels as a function of the number of unit cells.

nucleus and splitting will begin at small lattice spacings. The forbidden regions between two energy bands are called energy gaps or band gaps.

According to the Pauli principle, each atom will have two electrons in each level, so that there are $2N$ electrons to fit into the band. If each atom contributes two valence electrons, the band will be exactly full. This full band is called a 'valence band'. If the forbidden energy gap E_g between the highest filled band and the next empty band is large, no electron conduction can take place. The crystal behaves as an insulator. If, on the other hand, the value of E_g is small, some electrons from the lower filled band will be excited thermally into the upper empty band and conduction becomes possible. The upper empty band is therefore called the 'conduction band'. Materials may behave as semiconductors. When some electrons are excited into the conduction band, the resulting vacant states near the top of the filled band behave in every way as particles of positive charge $+e$. Such an empty state is called a 'positive hole'. If each atom contributes one valence electron to the band, the band will be half full and one might expect that the solid will possess the characteristic properties of a metal.

4.3.1.2 Optical properties of semiconductors [2, 101–106]

An electron is excited from the valence band to the conduction band with the absorption of a photon of energy approximately equal to the forbidden gap energy E_g. This is called the fundamental absorption. This process creates a free electron in the conduction band and leaves a hole in the valence band. For the simplest case of a direct-gap semiconductor, the absorption coefficient will be evaluated using parabolic $E-k$ curves centered about the wave vector $k = 0$, as shown in Fig. 4.8. Taking the zero of energy at the top of the valence band, the energy of an electron near the lower edge of the conduction band and an electron near the top of the valence band may be written, respectively, as

$$E_c(k) = E_g + (\hbar^2/2m_e)k^2 \tag{4.46}$$
$$E_v(k) = -(\hbar^2/2m_h)k^2. \tag{4.47}$$

Here E_g is the energy required to excite an electron from the top of the valence band to the bottom of the conduction band, m_e and m_h are the effective masses of the electrons in the conduction band and in the valence band (hole), respectively. For vertical transitions, the energy of a photon required to induce a transition is then given by

$$h\nu = E_c - E_v = E_g + \hbar^2 k^2 / 2m^* \tag{4.48}$$

where m^* is the reduced mass given by

$$1/m^* = (1/m_e) + (1/m_h). \tag{4.49}$$

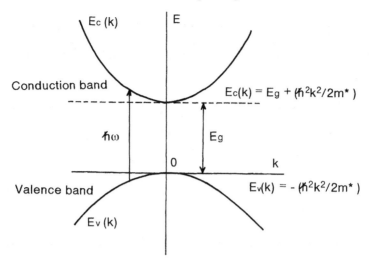

Fig. 4.8.   Plot of energy E versus wave vector k for an electron near the band gap and direct transitions. The vertical arrows indicate a variety of transitions that may be induced by optical radiation.

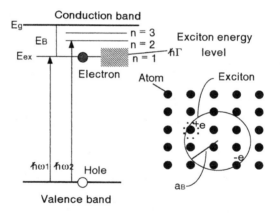

Fig. 4.9.   Exciton energy levels diagram.

If an electron and hole are present in reasonable proximity and relatively immobile, their Coulomb attraction may result in the bound state of an electron and hole, as shown in Fig. 4.9. Such a bound state is called an exciton [104–106]. It is analogous, in a sense, to the electric polarization of virtual excitation of an electron bound closely to an ion by Coulomb interaction (see Section 4.3). Therefore, an exciton can be described essentially as a collective electronic elementary excitation in a state of total wave vector K travelling through the insulator or semiconductor. The exciton energy is given by

$$E_{ex} = E_g - E_B + \hbar^2 K^2/2M^2 \qquad (4.50)$$

where E_g is the band gap energy, $\hbar^2 K^2/2M$ is the kinetic energy with which an exciton moves through the crystal, K is the wave vector of the center of mass, M is the translational mass ($= m_e + m_h$), and E_B is the exciton binding energy, which is given by

$$E_B = m^* e^4/2\hbar^2 \varepsilon^2 n^2. \qquad (4.51)$$

Here n is the principal quantum number. The binding energy for the lowest exciton state, $m^* e^4/2\hbar^2 \varepsilon^2$, is the energy required to separate an electron and hole in the 1s state. This energy is typically of the order of 20 meV in a bulk semiconductor. Because kT $\approx$ 26 meV at room temperature, low temperatures are usually necessary to prevent thermal dissociation of excitons [2]. The main experimental evidence for exciton formation comes from observations near the long wavelength edge of the main absorption band.

The average separation or effective radius of the lowest exciton state a_B is called the exciton Bohr radius and is estimated by [2]

$$a_B = \hbar^2 \varepsilon/m^* e^2$$
$$= (m_e/m^*)\varepsilon a_0 \qquad (4.52)$$

where m^* is the exciton-reduced mass defined in Eq. (4.49), ε is a static dielectric constant of the semiconductor and a_0 is the first Bohr orbit of the hydrogen atom ($= 0.53$ Å).

4.3.2 Linear optical properties of semiconductor-doped glasses

In a large semiconductor crystal, the overall shape and size of the crystal makes little or no difference to its original properties. As the crystal becomes smaller, however, the quantum confinement effect will split the bulk conduction (c) and valence (v) bands into a series of discrete energy levels. This situation can be drawn schematically as the opposite case to the band formation in Fig. 4.7(a). The effects of the surface also become increasingly important. These two phenomena make the optical properties of semiconductor microcrystallites very different from the bulk. Based on the classification made by A. L. Efros and A. L. Efros [107], and A. I. Ekimov *et al.* [108–110], the size effects are classified into two categories with a variation of the ratio of microcrystal radius R to exciton radius a_B: electron or hole confinement region and exciton confinement region.

If the microcrystallite radius is much smaller than the exciton Bohr radius ($R \ll a_B$), the energy of the size quantization of an electron is much larger than the energy of its Coulomb interaction with a hole, and the electron and hole are individually confined. The interband absorption spectrum should change from continuous bands to a set of discrete lines as the size of the crystallite is decreased, and probes the nature of the confinement of electron and hole state [107–110]. The confinement in the microcrystallite also leads to a blue shift of the position of the interband absorption edge. This is called the strong confinement limit, which has been observed in materials such as CdS and CdSe with large values of Bohr radius (bulk Bohr radius of CdSe is 30 Å). The quantity k in Eq. (4.46) can be represented in the form $k = \phi_{n,l}/R$ and the value of the shift is inversely proportional to the size of the microcrystals as expressed by [96, 107, 109–111]

$$E = E_{\mathrm{g}} + (\hbar^2/2m_{\mathrm{e}}R^2)\phi_{n,l}^2 \qquad (4.53)$$

where $\phi_{n,l}$ is a universal set of numbers independent of R. In the special case when $l = 0$ (s electron state), $\phi_{n,0} = n\pi$ ($n = 1, 2, \ldots$).

The quantum confinement effect in linear optical properties has been reported by many authors for various semiconductor-doped glasses. One example of the absorption spectra is shown in Fig. 4.10 for glasses doped with CdS_xSe_{1-x} microcrystallites that have various average radii [112]. The distinct discrete absorption lines as well as a blue shift of the energetically lowest

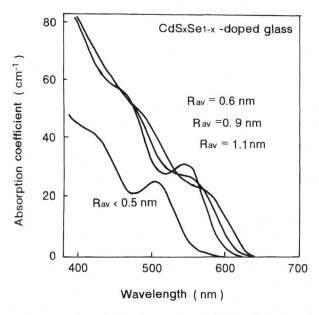

Fig. 4.10.   Absorption spectra of the glasses doped with CdS_xSe_{1-x} microcrystallites with various average radii. [Reprinted from H. Shinojima, J. Yumoto, N. Uesugi, S. Ohmi and Y. Asahara, *Appl. Phys. Lett.* **55** (1989) 1519, copyright (1989) with permission from the American Institute of Physics.]

electron-hole transition (absorption edge) are observed for each absorption spectrum.

Although the optical effects observed in the experimental CdS_xSe_{1-x}-doped glasses are due to electron and hole confinement, N. F. Borrelli *et al.* [113] emphasized that quantum confinement effects are not present in color filter glasses, because the crystals present in commercial glasses are too large (average diameter of about 100 Å) to display strong quantum size effects. According to their considerations, variations in optical properties are due to changes in the stoichiometry of the CdS_xSe_{1-x} during the crystallite growing process. A comparison of the selenium-to-sulfur mole fraction in the micro-crystallites obtained in an experimentally mixed anion glass, indicates a tendency for selenium to remain in the glass matrix, whereas sulfur is more easily incorporated in the microcrystallites. This may result from a large difference in diffusion coefficient between the S and Se ions. The shift in the energy of onset of absorption is consistent with the band-gap shift that varies with the selenium fraction within the CdS_xSe_{1-x} microcrystallites [113, 114].

In the other limiting case, if the microcrystal radius is much larger than the exciton Bohr radius in the material under investigation ($R \gg a_B$), a size quantization of all the excitons takes place [107, 108, 110]. When R is $2 \sim 4$ times a_B, an exciton forms whose center of mass motion is quantized and exhibits a discrete state within the crystallite. This is called the weak confinement limit. The spectral positions of the exciton absorption lines are given by [95, 107, 108, 110]

$$\hbar\omega = E_g - E_b + (\hbar^2\pi^2 n^2/2MR^2) \qquad (4.54)$$

where E_b is the exciton binding energy for the lowest 1s state in the bulk crystal. The maximum oscillator strength is exhibited by the transitions with $n = 1$. Then, the shorter-wavelength shift of exciton absorption and emission lines that is inversely proportional to the crystalline radius R is observed. The exciton confinement is expected in the glasses containing microcrystals of I–VII compound such as CuCl with a relatively small (~ 6 Å) exciton radius a_B [108]. In Fig. 4.11, a typical example of absorption spectra around Z_3 excitons in CuCl microcrystals is shown [17, 115]. The absorption peak of the Z_3 exciton shifts towards higher energies compared with the Z_3 exciton energy of bulk CuCl.

4.3.3 Nonlinear optical susceptibilities of the strong confinement case

In an early stage, optical nonlinear experiments in semiconductor-doped glasses were first made for a series of commercially available sharp-cut filter

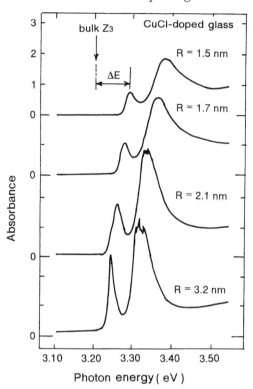

Fig. 4.11.   A typical example of absorption spectra around an exciton in CuCl doped glasses at 77 K. The arrow indicates the Z_3 exciton energy of the bulk crystal. [Reprinted from T. Tokizaki, T. Kataoka, A. Nakamura, N. Sugimoto and T. Manabe, *Jpn. J. Appl. Phys.* **32** (1993) L782, copyright (1993) with permission from Publication Office, Japan Journal of Applied Physics.]

glasses [14] consisting of CdS_xSe_{1-x} microcrystals (typically $100 \sim 1000$ Å in size) suspended in a glass matrix ($0.3 < x < 0.8$). A nonlinear response in the nanosecond range and large nonlinear susceptibilities of the order of 10^{-8} esu have been observed in this experiment. Table 4.5 lists the nonlinear characteristics of sharp-cut filter glasses reported to date in a number of papers.

Various kinds of nonlinear optical processes related to the population of real states have been used to explain the nonlinearities near the band edge of a direct-gap semiconductor. In small direct-gap semiconductors, the band filling effect has been identified as the primary mechanism for a nonlinear optical effect [2, 10, 98]. When the semiconductor material is pumped near the band gap, a high-density electron-hole plasma is generated at room temperatures. They eventually lose their energy and relax to the bottom of the conduction band mainly by emitting phonons. This results in the filling of the states near

Table 4.5. *Nonlinear characteristics of CdS$_x$Se$_{1-x}$-doped sharp cut filter glasses*

Sample No.	Measurement	Wavelength λ (nm)	Nonlinear susceptibility $|\chi^{(3)}|$(esu)	Response τ (ps)	Ref.
CS2–62	PP	566.5		~400	116
CS2–77	TBSD		2.1×10^{-9}		117
CS3–69	NFWM	560, 590	6.9×10^{-12}		117
CS3–69	KS			~3.6	117
CS3–69	DFWM	532		90	120
CS3–68	DFWM	532		~80	121
CS3–68	DFWM	532	$\sim1.3 \times 10^{-8}$		14
CS2–73	DFWM	580	$\sim5 \times 10^{-9}$		14
CS2–62	DFWM		8.4×10^{-11}		131
CS2–73	DFWM		2.6×10^{-10}		131
CS3–66	DFWM		2×10^{-10}		131
CS3–68	DFWM		1.6×10^{-10}		131
OG570	PP	580 ~ 590	8×10^{-8}	~30	119
OG570	DFWM	532		1000	118
OG570	DFWM	532	$\sim10^{-8}$		122
OG590	DFWM	580		~15	123
OG590	DFWM, NFWM	589	10^{-7}		124
OG590	DFWM	584	2.5×10^{-8}		125
OG530	DFWM	532	7×10^{-12}		128, 130
RG695	DFWM	694.5	$\sim3 \times 10^{-9}$	<8000	14
RG630	DFWM		$\sim10^{-7}$		126
RG630	DFWM (125K)		$\sim10^{-4}$ (125K)		126
RG930, 715	MQH	620		~100	127
Y–52	BS	532	1.3×10^{-9}	25	129
RG715	NFWM	–	–	10	132
OG530	DFWM	532		~50	133
CS3–68	DFWM	532		~100	133

PP, pump-probe technique; DFWM, degenerate four-wave mixing; NFWM, non-degenerate four-wave mixing; KS, Kerr shutter technique; BS, Bistable switching; TBSD, two-beam-self-diffraction.

the bottom of the conduction band with electrons, preventing further absorption into these states by Pauli exclusion. This effect is often referred to as a 'blocking effect' in the sense that the filling effect blocks absorption states near the band edge. The absorption saturates in the vicinity of the band edge, which shifts the band edge to higher energies with increasing laser intensity.

Although it is not perfect, the band-filling model described above in bulk semiconductors leads to a good qualitative understanding of the nonlinear properties of semiconductor-doped glasses [122]. As discussed in the previous Section 4.3.3, the absorption spectrum changes to a series of discrete lines as the microcrystallite size decreases. This quantization is accompanied by concentration of the bulk oscillator strength into a single discrete line, leading to a further enhancement of this nonlinearity. Therefore, the process is often referred to as 'state-filling' in the microcrystallite as opposed to 'band-filling' in the bulk semiconductor [122, 134]. The change in absorption $\Delta\alpha$ gives the nonlinear refractive index changes Δn, through the Kramers–Kronig transformation [11, 12].

Figure 4.12 shows the measured change of absorption coefficient $\Delta\alpha(\lambda)$ (solid line) and the dispersion of the nonlinear refractive index $n_2(\lambda)$ for CdS_xSe_{1-x}-doped glass [11, 12]. The dashed curve shows the computer-generated $n_2(\lambda)$ dispersion by Kramers–Kronig transformation of the measured $\Delta\alpha(\lambda)$ and gives n_2 as a function of wavelength. The value of n_2 is negative below and positive above the band gap.

The third order nonlinear susceptibility is given by [135]

$$\chi^{(3)} = 6p|P_{cv}|^4/4\pi R^3\hbar^3(\omega - \omega_1)^3 \qquad (4.55)$$

where P_{cv} is the bulk transition dipole moment, ω_1 is the resonant frequency of the transition to the lowest state, p is the volume fraction of microcrystallite in the glass matrix and R is the crystallite radius. Consequently $\chi^{(3)}$ will be proportional to the reciprocal of the confinement volume and will increase with

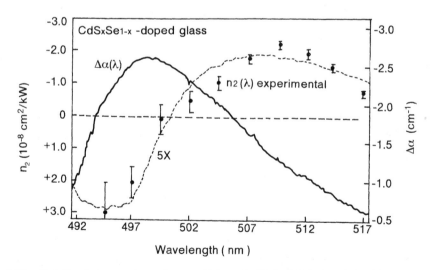

Fig. 4.12.   Relation between absorption coefficient $\Delta\alpha(\lambda)$ (solid line) and the dispersion of the nonlinear refractive index $n_2(\lambda)$ for CdS_xSe_{1-x}-doped glass. [Reprinted from G. R. Olbright and N. Peyghambarian, *Appl. Phys. Lett.* **48** (1986) 1184, copyright (1986) with permission from the American Institute of Physics.]

decreasing R. This expression for the nonlinear susceptibility is applicable particularly to cases in which the II–VI family semiconductors are doped in the glass matrix. The enhancement of the nonlinearity due to quantum confinement has been confirmed experimentally [134, 136]. The variation of $|\chi^{(3)}|$ of a glass that was doped with CdS_xSe_{1-x}-crystallite with mean particle size $(24 \sim 30 \text{ Å})$ versus wavelength are shown in Fig. 4.13 [134]. Higher efficiencies, i.e. larger nonlinearities, are obtained for glasses containing smaller particles. These observations are in agreement with the theoretical prediction. The ratio $\chi^{(3)}/\alpha$ was about three times greater for a radius of 24 Å than for a 30 Å radius. One of the reasons for this is that the nonlinear susceptibility is only due to the 1s-1s transition, increasing as $1/R^3$, whereas the absorption coefficient increases less rapidly since higher energy transitions are more off resonance [136].

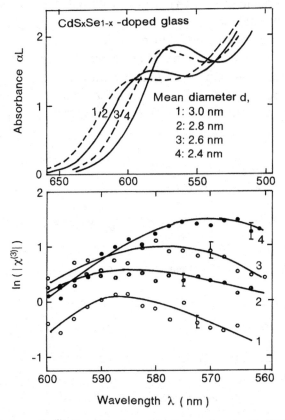

Fig. 4.13.   Variation of $|\chi^{(3)}|$ (logarithmic scale) of glass doped with CdS_xSe_{1-x}-crystallite with mean particle size $(24 \sim 30 \text{ A})$ versus wavelength. [Reprinted from D. Richard, P. Rousignol, F. Hache and Ch. Flytzanis, *Phys. Stat. Sol. (b)* **159** (1990) 275, copyright (1990) with permission from Wiley VCH Verlag Berlin GmbH.]

As shown in Fig. 4.14, an increase of $|\chi^{(3)}|$ (logarithmic scale) with increasing volume fraction of microcrystallite was also reported for CdSe-doped glasses [137, 138] that were fabricated using the super-cooling technique. A third-order nonlinear susceptibility of the order of 10^{-6} esu was obtained for these glasses.

4.3.4 Nonlinear response time in the strong confinement case

4.3.4.1 Three-level system

As previously discussed in Section 4.3.3, state filling has been identified as the primary mechanism for the observed optical nonlinearity in semiconductor-doped glasses in the strong confinement case. Such a simple model, however, is not always able to explain quantitatively the nonlinear characteristics observed, such as the size dependence [140, 141] or intensity dependence [139] of $|\chi^{(3)}|$. As the magnitude of the state filling depends on the density of the photo-induced carriers in the conduction band regardless of the excitation mechanism, the nonlinearity must clearly be affected by the carrier recombination dynamics. We should also remember that the model does not take carriers [142] that are trapped in the defects or surface states into account. They especially play an important role when microcrystallite size is small. The

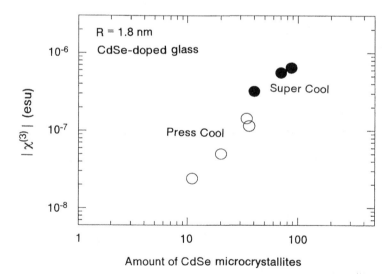

Fig. 4.14. The third-order nonlinear susceptibility $|\chi^{(3)}|$ (logalithmic scale) as a function of CdSe microcrystallite concentration (arbitrary unit). [Reprinted from Y. Asahara, *Advances in Science and Technology* **4**, *New Horizons for Materials*, ed. P. Vincenzini, (1995) 91, copyright (1995) with permission from Techna Publishers Srl.]

smaller the size of microcrystallites, the more important the surface states become since the surface to volume ratio is larger for small particles. Furthermore, the other effects such as photodarkening and the intensity-dependent response also make the phenomena more complex [133].

In optical phase conjugation studies or pump-probe experiments, the decay curve or recovery time of adsorption saturation decomposes the curve into the sum of exponential components with a short time constant of ps and with a large time constant of ns [116, 118, 121, 127, 139, 143, 144]. This means that the photo-excited carriers in semiconductor-doped glasses have fast and slow relaxation mechanisms. The fast decay processes are assigned to the relaxation of free carriers down to the valence band or nonradiative relaxation to the trap level, whereas the slow decay processes are thought to be related to the slow relaxation of the carriers trapped in defects that predominantly exist in the microcrystallites.

As has been reported in several papers, the energy level diagram proposed for semiconductor-doped glasses is, therefore, simplified to a three-level system including the valence band, the conduction band and the trap levels, as shown in Fig. 4.15 (a).

The existence of the trap levels is known by the fact that the semiconductor-doped glasses show very broad photoluminescence spectra caused by the

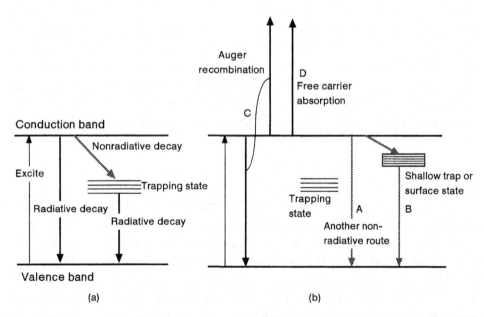

Fig. 4.15. Energy level diagram of glasses doped with semiconductor microcrystallites with various relaxation process, (a) three-level system, (b) modified three-level system.

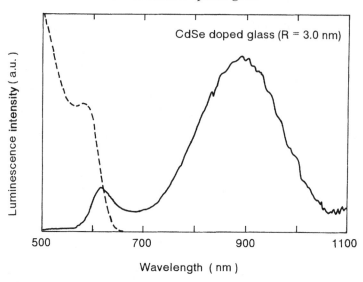

Fig. 4.16.   Photoluminescence   spectrum   of   CdSe-doped   glass.   ($R = 3.0$ nm)   (50ZnO-50P$_2$O$_5$ + 4CdSe (mol %)). [Reprinted from Y. Asahara, *Advances in Science and Technology* **4**, *New Horizons for Materials*, ed. P. Vincenzini, (1995) 91, copyright (1995) with permission from Techna Publishers Srl.]

relaxation of trapped electrons in the near infrared region [137, 138, 145, 148]. As shown in Fig. 4.16, for instance, the photoluminescence spectrum of a CdSe-doped glass consists of a narrow band near the absorption edge and a broad band in the longer wavelength region [138], which are respectively assigned to the direct recombination of the excited carriers to the valence band and to the radiative recombination via the trap levels [145]. If some electrons can be trapped, the remaining electrons in the conduction band can still contribute to the band filling. When the intensity is increased, the trap becomes occupied and thus unavailable. Then the absorption is thought to start saturating, giving rise to a resonant $\chi^{(3)}$ [139].

4.3.4.2 Intensity-dependent response

According to the simple models, the absorption coefficient asymptotically approaches zero as the states are filled. The response of semiconductor-doped glasses has been found, however, to saturate to lower intensity than that normally expected from simple models of the nonlinear response [120, 126, 133, 139], i.e. the absorption coefficient does not completely saturate to zero [147–149]. When the pulse intensity increases, the phase conjugate reflectivity finally levels off the relationship, for instance, at a fluence of $\sim$1 mJ/cm^2 for which the reflectivity is proportional to I^2, as shown in Fig. 4.17.

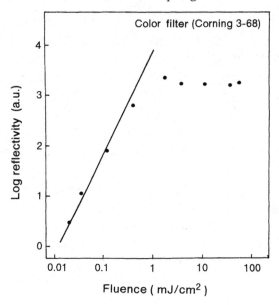

Fig. 4.17. Log-log plot of the phase conjugation reflectivity as a function of the pump-beams energy fluence for sharp cut filter glass (Corning 3-68). [Reprinted from P. Roussignol, D. Richard, J. Lukasik and C. Flitzanis, *J. Opt. Soc. Am.* **B4** (1987) 5, copyright (1987) with permission from the Optical Society of America.]

This nonsaturable part of the absorption, or saturation of the nonlinear effect, has been interpreted by the intensity-dependent response. Based on the results of a number of nonlinear response and photoluminescence decay experiments, the intensity-dependent response can be divided phenomenologically into two categories: reversible or irreversible. The latter is known as a darkening effect and drastically reduces the response time and the nonlinearity, as will be discussed in more detail in Section 4.3.5.

There has been evidence for a response time that is reversibly dependent on the pump intensity [118, 150]. On the basis of the dependence of the photoluminescence spectra of semiconductor-doped glasses on the pump laser intensity [151–153], various reversible mechanisms which determine the relaxation times at high intensity were discussed. A modified three-level model that involves the valence band, the conduction band and several kinds of intermediate trap levels was proposed to explain the intensity-dependence of the response time. Schematic representations of the electronic levels of the suggested model are shown in Fig. 4.15. At low pump intensities, trap recombination dominates and results in long lifetime. As the intensity of the pumps is increased, the traps begin to fill. As soon as the traps become occupied, the trapping becomes less efficient, and the electrons choose another way to recombine, giving rise to direct recombination (Fig. 4.15(b) A) [151, 152] or recombination through different groups of traps such as shallow traps or surface states which (Fig. 4.15(b) B) [150, 153–156] then contribute to the fast response. In the latter case, the filled traps are thought to lie in the volume of the particles [154].

As shown in Fig. 4.15(b) C and D, the other possible relaxation mechanisms which can interpret the nonsaturable absorption are Auger recombination of the electrons and the holes [118, 132, 157, 158] or free-carrier absorption [147, 159], both of which lead to a decrease of the excited carrier lifetime and prevent band filling. K. W. Delong *et al.* [160], however, comment that free-carrier absorption is not significant in these systems. The saturation of the nonlinear index change due to the thermal effect [18, 148], or some more complex mechanisms [161, 162] have been proposed recently.

4.3.5 Photodarkening effect

4.3.5.1 Photodarkening mechanisms

One further phenomenon worth mentioning in connection with nonlinear optical properties is the photodarkening effect. The long radiative recombination times of semiconductor-doped glasses in the strong confinement case still restricts their use in high-speed applications. The photodarkening effect, and therefore an understanding of the processes that cause this effects is important particularly in reducing the response times and has been studied using various techniques, such as absorption saturation, induced-absorption, luminescence, and degenerate four-wave mixing.

The photodarkening effect is a permanent effect after a long period of intense laser irradiation [133]. This effect appears with a decrease in both the nonlinear optical response time [133, 163] and the nonlinear efficiency such as the diffraction efficiency [133, 163, 164, 167] or absorption saturation [133, 159, 160], which sometimes accompanies permanent grating [133, 165, 166] or darkening [133, 147, 159, 160, 168]. Photodarkening also significantly affects the photoluminescence characteristics [133, 137, 138, 147, 152, 159, 160, 164, 169, 170]. A drastic reduction of the lifetime and luminescence efficiency of the direct recombination of photo-induced carriers has been observed with a reduction of the broad luminescence of deeply trapped carriers. The changes are permanent at room temperature but a darkened glass recovers its original properties when it is heat-treated at approximately 450 °C for several hours [133].

Typical experimental results are shown in Fig. 4.18 [163]. The decay time of a nonlinear grating signal is of the order of 1ns immediately after short laser exposure. This component is, however, reduced dramatically with exposure time and finally a fast decay, as short as 7 ps, remains after a 3 h exposure. No darkening was observed, so the process is referred to as a 'photo annealing effect' in this experiment. This annealing process is favorable from the viewpoint of optical devices, since a reduction of response time is dominant over the reduction of nonlinearity, i.e. the signal peak intensity becomes one-third after annealing, whereas the speed (decay time) is enhanced ∼100 times. Figure 4.19 also shows one example of the change in the photoluminescence decay characteristics upon intense laser irradiation [137].

Since the undoped host glass showed none of these effects under the same illumination conditions, it appears that the darkening effect is not due to the glass host matrix alone: other factors such as solarization effects which are known to occur in optical-grade glasses under intense laser radiation may not be involved. Of course, such darkening effects can not be seen in bulk semiconductors either, so it is apparent that this effect is due to either the particular structure of the crystallites or the semiconductor-glass interface at the surface of the microcrystallites [160].

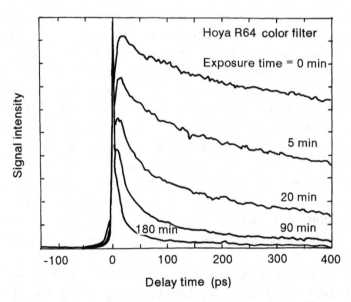

Fig. 4.18. Signal intensity versus probe-pulse delay time for various exposure times. (CdS$_x$Se$_{1-x}$ filter glass, $\lambda = 613$ nm, pump power: 1 mW.) [Reprinted from M. Mitsunaga, H. Shinojima and K. Kubodera, *J. Opt. Soc. Am.* **B5** (1988) 1448, copyright (1988) with permission from the Optical Society of America.]

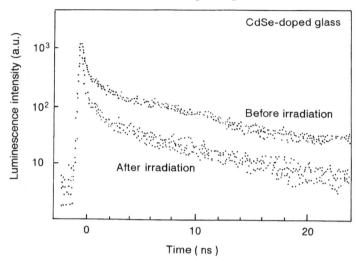

Fig. 4.19. Photoluminescence decay curves of CdSe-doped glass before and after 30 min laser irradiation (1 MW/cm²). [Reprinted from Y. Asahara, *Advances in Science and Technology* **4**, *New Horizons for Materials*, ed. P. Vincenzini, (1995) 91, copyright (1995) with permission from Techna Publishers Srl.]

One possible mechanism of this effect is the intensity dependence of the response time, such as free carrier absorption [159] as described in Section 4.3.4. This effect, however, can be eliminated as an origin of the photodarkening effect because the explanation requires a permanent decrease of recombination time, which contradicts the experimental results. The temperature rise upon irradiation is calculated to be too small for any thermally induced change in the material [168], which rules out any thermal mechanism [133]. The darkening effects are significantly reduced at low temperature [169], suggesting that the process includes some form of thermal activation across a potential barrier. In addition, the accompanying photoluminescence spectral changes suggest the presence of trap levels within the gap or the process that give rise to a change within the trapping state that seems to play an important role in the darkening effect. It was supposed at first that trapping sites are no longer available or trapping is reduced in the process of intense laser exposure [133], while a new nonradiative decay route of the carriers at the band edge state is induced [168].

On the basis of the experimental results, several models that are shown in Fig. 4.20 have been proposed for the photodarkening process. One model is the modified three-level model consisting of two types of deep trap levels [164, 169, 171, 172]: the same trap level T which is responsible for the luminescence and phase conjugation process, and a different kind of trap T' or surface trap

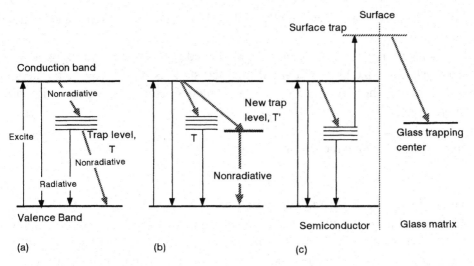

Fig. 4.20. Various mechanisms proposed for explaining photodarkening effect.

which is created under the influence of laser exposure is used to open an additional channel for nonradiative recombination of the excited carriers. Excited carriers in the conduction band will either undergo direct recombination, be responsible for the higher-energy luminescence level or will populate initially only trap sites of the kind T that are already present in the semiconductor microcrystallites. Then, the carrier will populate trap level T', the number of concentrations of which increases as a function of laser exposure. This situation is illustrated in Fig. 4.20 (b).

The second modified model [137, 138, 170] includes an additional opening for a fast new nonradiative recombination route of carriers in the trapping state, i.e. the growth of the number of electon states on the surface by photo-darkening as a cause of the rising decay rate of the carriers. The new recombination route is probably a nonradiative one. This leads to a shorter conduction-band carrier lifetime and a smaller luminescence efficiency in the darkened sample. This is illustrated in Fig. 4.20 (a).

The third model is based on the ionization of microcrystals under intense irradiation [168, 173] by means of a two-step excitation process, which is illustrated in Fig. 4.20 (c). Upon irradiation, electrons in the valence band are excited to the conduction band, some of which relax to trap levels at the glass–semiconductor interface. The key idea of this explanation is that before the electrons can recombine, some of those trapped electrons are re-excited to higher-energy surface states, from which they may migrate into the bulk of the glass. These electrons are eventually captured by the deep traps in the structure of the glass.

The relaxation dynamics of the photodarkened glass depends strongly on the microcrystalline size [173]. In small microcrystallites, fast decay components are dominant. The decay curve for a semiconductor-doped glass with an average crystalline radius of 10 Å consisted almost completely of a fast decay component with 4 ps time constant, which is assumed to be due to carrier recombination through the surface states.

4.3.5.2 Origin of the trap levels

Several groups have attempted to determine the origin of the trap levels, existing predominantly in the semiconductor-doped glasses or induced by the intense laser irradiation. As for the origin of the trap levels that predominantly exist in CdS- or CdSe-doped glasses, the possibilities of impurities (zinc, etc.), defects owing to sulfur or selenium vacancies, or sufficient Cd^{2+} have been proposed [113, 146, 151, 171]. It is well known that sulfur has a high vapor pressure and can easily oxidize. Therefore, some amount of sulfur may be lost in the process of glass melting, and the S to Cd ratio in the glass must be less than that in the batch, which may result in the trap center becoming a sulfur-deficient center or Cd-sufficient center which may be on the CdS microcrystallite surface [174].

Then, what are the photo-induced trap levels? It was reported that CdTe-doped glass eliminates the presence of trap centers [137, 138, 175, 176]. The photoluminescence spectrum of this glass consists of only one narrow peak near the band edge, as shown in Fig. 4.21 [138]. Figure 4.22 shows that the fast decay element due to the nonradiative decay observed in the CdSe-doped glass also disappeared in the CdTe-doped glass [138]. Another interesting phenomenon is that the photodarkening effect cannot be observed upon intense laser irradiation in this glass [138], as shown in Fig. 4.23. It has been also shown that the H_2O-treated [146] or hydrogenated [177] CdSe-doped sample seems to be more resistant with respect to the darkening effect. When treated with H_2O, it probably diffuses into the glass matrix as OH^- and neutralizes the trapping sites such as the vacant Se sites at the microcrystallite surface [146]. All of the results described above suggest that the photodarkening effect needs 'trap levels' that already exist in the glasses and could well be interpreted as the creation of a new channel for nonradiative decay of carriers from these 'trap levels'.

One possibility is the buildup of efficient recombination centers by photo-formation of Cd^{2+} in the crystallites [147]. Another possibility is associated with the state of surface activation due to the production of a layer of impurity

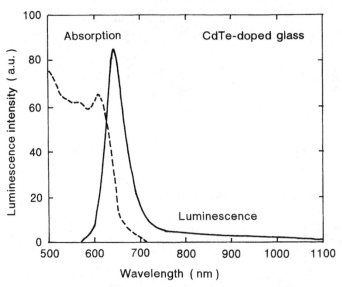

Fig. 4.21.   Photoluminescence spectra of CdTe-doped glass. [Reprinted from S. Omi, H. Hiraga, K. Uchida, C. Hata, Y. Asahara, A. J. Ikushima, *et al. Proc. International Conference on Science and Technology of New Glasses*, ed. S. Sakka and N. Soga (1991) p. 181, copyright (1991) with permission of The Ceramic Society of Japan.]

ions around the semiconductor microcrystallites. Apparent grain growth of microcrystallite, which may act as an interface between microcrystallites and the glass structures has been confirmed experimentally [172, 178].

4.3.6 *Nonlinear optical properties in the weak confinement case*

4.3.6.1 *Excitonic enhancement of $\chi^{(3)}$ in microcrystallite*

E. Hanamura *et al.* [95, 135, 179] have discussed the nonlinear optical properties in the weak confinement case in which the size of microcrystallites is much larger than the bulk exciton. The excitons produce much weaker nonlinear optical effects than electron-hole plasma in bulk materials. In bulk materials, the excitons are stable at low temperatures, but are unstable against collisions with energetic phonons at temperatures above room temperature [2, 97].

Much higher nonlinearities can be theoretically predicted for excitons in the weakly confined system [95, 135, 179]. When a crystallite radius R is relatively small but larger than an effective Bohr radius a_B of a exciton ($a_B \ll R < \lambda$), the character of an exciton as a quasiparticle is well conserved at relatively high temperature and the translational motion of the exciton is confined to

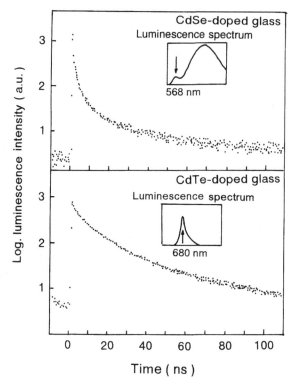

Fig. 4.22. Photoluminescence decay curves of CdSe and CdTe doped glasses. [Reprinted from S. Omi, H. Hiraga, K. Uchida, C. Hata, Y. Asahara, A. J. Ikushima, *et al. Proc. International Conference on Science and Technology of New Glasses*, ed. S. Sakka and N. Soga (1991) p. 181, copyright (1991) with permission of The Ceramic Society of Japan.]

microcrystallites. The transition dipole moment to this exciton state from the crystal ground state is given by

$$P_n = (2(2)^{1/2}/\pi)(R/a_B)^{3/2}(1/n)P_{cv}. \tag{4.56}$$

The confinement of the excitonic envelope wavefunction is expected to give rise to enhancement of the transition dipole moment by a factor of $(R/a_B)^{3/2}$ in comparison with the band-to-band transition P_{cv}. In addition, the oscillator strength is almost concentrated on the lowest exciton state, i.e. the principal quantum number for the center-of-mass motion $n = 1$. Such a giant-oscillator-strength effect will result in the enhancement of the nonlinear polarizability under nearly resonant excitation of excitons in the weak-confinement case. The large interaction energy between two excitons, i.e. the large oscillator strength of a biexciton also may lead to a further enhancement of nonlinearities. Under a slightly off-resonant pumping of the lowest exciton $E_1 = \hbar\omega_1$, and

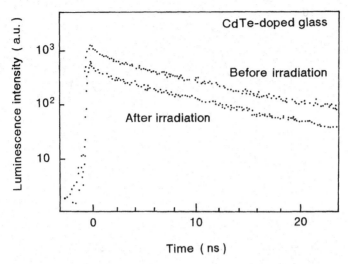

Fig. 4.23. Photoluminescence decay curves of CdTe-doped glasses before and after 30 min irradiation. [Reprinted from H. Hiraga, C. Hata, K. Uchida, S. Ohmi, Y. Asahara, A. J. Ikushima, *et al. Proc. XVI International Congress on Glass*, vol. **3** Madrid (1992) p. 421, copyright (1992) with permission of Sociedad Espanola de Ceramica y Vidrio.]

$\omega_{\text{int}} > |\omega - \omega_1| > \Gamma$, the third-order nonlinear susceptibility $\chi^{(3)}$ is given by [95, 135, 179]

$$\chi^{(3)} \sim N_{\text{c}} |P_1|^4 / \hbar^3 (\omega - \omega_1)^3 \sim (R/a_{\text{B}})^3 \qquad (4.57)$$

where N_{c} is the number density of microcrystallites per unit volume ($= 3p/(4\pi R^3)$) with p a constant volume fraction, ω_1 is the resonant frequency of the lowest exciton, Γ is the transverse relaxation, $\hbar\omega_{\text{int}}$ is the interaction energy of the two excitons which makes them deviate from harmonic oscillators given by [135, 180]

$$\hbar\omega_{\text{int}} = (13/4)(a_{\text{B}}/R)^3 E_{\text{b}}. \qquad (4.58)$$

For this case, the dipole moment P given by Eq. (4.56) is multiplied four times in the expression for the third optical nonlinearity, so that $\chi^{(3)}$ is almost real and increases as R^3 because the fourth power of P_1 gives R^6-dependence and overcomes the R^{-3} dependence of N_{c}.

Under a resonant excitation and $\omega_{\text{int}} > \Gamma > |\omega - \omega_1|$, $\chi^{(3)}$ is almost purely imaginary and contributes to the absorption saturation. In this case, creation of resonant excitons at room temperature should produce a weak bleaching of the exciton resonances caused by the phase-space filling and exchange effect [178]. When Γ and γ are independent of the size R, Im $\chi^{(3)}$ is also proportional to R^3. The estimation of the dependence of the third-order optical susceptibility $\chi^{(3)}$ on the radius of CuCl microcrystallite is shown in Fig. 4.24 (a) [95].

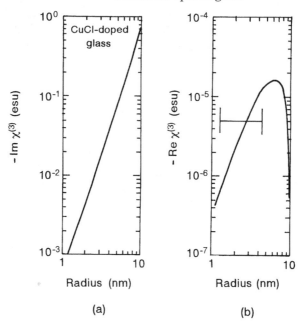

Fig. 4.24.   The dependence of the third-order optical susceptibility $\chi^{(3)}$ on the radius of CuCl microcrystallite. (a) $-\text{Im}\,\chi^{(3)}$ at $\hbar(\omega_1 - \omega) = 0$ meV, (b) $-\text{Re}\,\chi^{(3)}$ at $\hbar(\omega_1 - \omega) = 2.5$ meV. The bar indicates the enhancement region. [Reprinted from E. Hanamura, *Optical and Quantum Electron.* **21** (1989) 441, copyright (1989) with permission from Chapman and Hall Ltd.]

Another effect of the enhancement of the transition dipole moment is rapid super-radiative decay of the excitons as long as the radius R is less than the wavelength (λ) of the relevant light. The exciton has the maximum cooperation number in the sense that this exciton state is made by a coherent superposition of all atomic excitations which make up the microcrystallites. The super-radiative decay rate of the lowest exciton 2γ is evaluated as [95, 179]

$$2\gamma = 1/\tau = 64\pi(R/a_{\text{B}})^3(4P_{\text{cv}}^2/3\hbar\lambda^3) \tag{4.59}$$

where $\lambda = 2\pi c\hbar/E_1$. The excitons in the larger microcrystallite can radiate more rapidly with decay time $\tau \sim R^{-3}$. The lowest exciton in the CuCl microcrystallite (for example, with a radius of 5 nm) is estimated to have a decay time of 0.1 ns. In bulk semiconductors, the exciton absorption spectrum is thought to be strongly modified by the plasma screening effect [97, 181]. In the three-dimensionally confined microcrystallites, the allowed states are separated by large energies so that such transitions are difficult and screening is not thought to be so important [97, 98].

Under the confinement condition described above, the exciton can contribute to the enhancement of the nonlinear optical response and $\chi^{(3)}$ increases as R^3.

Three factors, however, can make the excitons in microcrystallite deviate from this condition [95, 135]. When R is further increased, first the exciton–exciton interaction $\hbar\omega_{int}$ is inversely proportional to the volume of the microcrystallite, according to Eq. (4.58). Then, if we assume $\omega_{int} > |\omega - \omega_1|$, the enhancement limit is given by

$$R < \{13E_b/4|\omega - \omega_1|\}^{1/3} a_B. \tag{4.60}$$

Since quantization energy is given by

$$E = n^2 \pi^2 \hbar^2 / 2MR^2 \tag{4.61}$$

the energy spacing is expected to decrease as the radius R of the microcrystallite increases. When the off-resonant energy $\hbar |\omega - \omega_1|$ becomes larger than the quantization energy $\hbar(\omega_2 - \omega_1) = 3\hbar^2\pi^2/2MR^2$, the lowest excitation level is hybridized with higher levels with the smaller transition dipole moments, which reduces the enhancement of $\chi^{(3)}$. Therefore, the enhancement limit is given by

$$R < [3\hbar^2\pi^2/2M|\omega - \omega_1|]^{1/2}, \tag{4.62}$$

When the size of the microcrystallite R increases beyond the coherent length of the exciton $L^* = (\hbar/MT)^{1/2}$, or the broadening $\hbar\Gamma$ becomes larger than the quantization energy, the lowest exciton level is also hybridized with higher levels that have smaller transition dipole moments. This also reduces the enhancement of $|\chi^{(3)}|$ [95, 135, 180] and if we assume $R < L^*$, the enhancement limit is given by

$$R < \{9(2\pi)^{1/2}/8\}^{1/3}(\hbar/M\gamma)^{1/2} \tag{4.63}$$

where 2γ is the longitudinal decay rate. From the limit of the three factors described above, an optimum size of the microcrystallite in order to get the largest optical nonlinearity may be expected, as shown in Fig. 4.24 (b).

4.3.6.2 Excitonic optical nonlinearity in semiconductor-doped glasses

A. Nakamura *et al.* [17, 115] observed large nonlinear susceptibilities under nearly resonant pumping of the exciton in CuCl-doped glasses. As shown in Fig. 4.25, the $\chi^{(3)}$ measured at Z_3 exciton energy of CuCl microcrystallites increases with an increase of R and subsequently decreases after reaching a maximum value at ~ 5.5 nm. The radius dependence is approximately R^3. The highest value is $|\chi^{(3)}| = 3.4 \times 10^{-6}$ esu for $R \sim 5.5$ nm. B. L. Justus *et al.* [182] also found that the excitonic optical nonlinearity in CuCl-doped borosilicate glass increases with increasing particle radius over the range 22 to 34 Å, in agreement with theoretical predictions.

A considerable light-induced blue shift of the excitonic band and absorption saturation were observed for CuCl crystallites embedded in a glass matrix

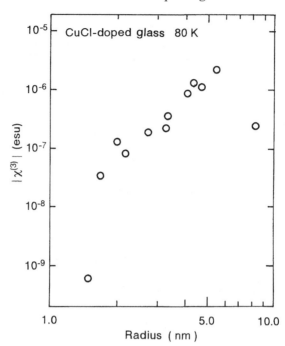

Fig. 4.25.   The size dependence of $\chi^{(3)}$ measured at Z_3 exciton energy of CuCl microcrystallites at 80 K. [Reprinted from T. Tokizaki, T. Kataoka, A. Nakamura, N. Sugimoto and T. Manabe, *Jpn. J. Appl. Phys.* **32** (1993) L782, copyright (1993) with permission from Publication Office, Japan Journal of Applied Physics.]

[183–185]. An investigation was also made of the linear and nonlinear optical properties of CuBr microcrystallites embedded in glass [186, 187]. In this case, at higher intensities of excitation, a strong blue shift and a bleaching is measured even at 300 K [187].

The radiative decay rate of the confined exciton in CuCl microcrystallites with radii R of $18 \sim 77$ Å embedded in glass was reported by A. Nakamura *et al.* [188]. They observed that the radiative decay rate is proportional to $R^{-2.1}$ as shown in Fig. 4.26, which confirms the theoretical prediction for the size dependence of the radiative decay rate of confined excitons. The slope is, however, smaller than the theoretically expected R^{-3} dependence from Eq. (4.59). They suggested that the deviation may come from the reduction of confinement for the smaller radius due to the difference in the dielectric constants of semiconductor microcrystallite and glass matrix.

From the viewpoint of applications, the figure of merit $\chi^{(3)}/\alpha\tau$ is of interest. As is expected from the results of Fig. 4.25 and 4.26, the figure of merit is increased with an increase of radii for $R < 4$ nm [115, 189]. The size

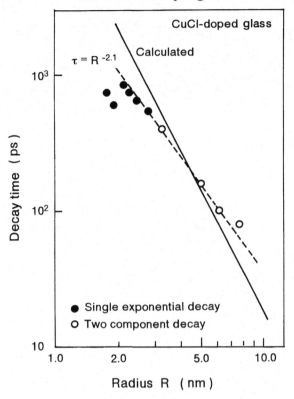

Fig. 4.26. The decay times of the Z_3 excitons as a function of the crystallite radius R. [Reprinted from A. Nakamura, H. Yamada and T. Tokizaki, *Phys. Rev.* **B40** (1989-II) 8585, copyright (1989) with permission from The American Physical Society.]

dependence of optical nonlinearity in $CuBr_xCl_{1-x}$ ($R = 2.7 \sim 56$ nm) and in CuBr ($R = 1.7 \sim 42$ nm) nanocrystals embedded in borosilicate glass have been also investigated [190–192]. The size dependence of $\chi^{(3)}/\alpha\tau$ for $CuBr_xCl_{1-x}$ exhibits an $R^{2.1}$ dependence and does not saturate up to the largest size of $R \sim 34a_B$. This behaviour is different from that observed in CuCl microcrystal, which has been attributed to the valence band structure of CuBr. This extremely high performance makes the CuCl- and CuBr-doped glasses almost ideal nonlinear optical materials at low temperature. However, this situation can only be achieved at temperatures that are low enough to prevent population of the higher-lying center of mass states.

4.3.7 Fabrication techniques

The commercial semiconductor-doped filter glasses are normally manufactured by quenching appropriately doped melts, and reheating them just above the

Table 4.6. *Nonlinear optical properties of semiconductor-doped glasses prepared by various fabrication technologies*

Dopant	Diam. d (Å)	Matrix	Process	Measurement	Absorption Coeff. α(cm^{-1})	Wavelength λ (nm)	Nonlinear susceptibility $|\chi^{(3)}|$ (esu)	n_2^{π} (cm^2/W)	τ (ps)	Ref.
CdS	35 ~ 60	S	SG	TBSD	~1715	460	$\sim6.3\times10^{-7}$			195
CdS$_x$Se$_{1-x}$	20 ~ 200	BS	MQH	TBSD		580 ~ 620			4 ~ 90	173
CdS$_x$Se$_{1-x}$	440 ~ 2400	P	MQH	Z		1064		2.5×10^{-13}		198
CdS$_x$Se$_{1-x}$		FP	MQH	DFWM		584	1×10^{-7}		14000	125
CdS$_x$Se$_{1-x}$ (x = 0.41)		FP	MQH	FWM		587	$\sim10^{-7}$			130
CdS$_x$Se$_{1-x}$ (x = 0.73)		FP		FWM		532	$\sim2.9\times10^{-9}$			134
CdSe	15	S	SPT	DFWM	1270 (600 nm)	532	1.3×10^{-8}		10 ~ 60	194
CdSe	~25			DFWM		546	$0.3\sim2\times10^{-9}$			165
CdSe	25 ~ 40	P	MQH	TBSD	526		$\sim1\times10^{-6}$			200
CdSe	~70	S	IBS	TBSD		445 ~ 510	1×10^{-8}			201
CdTe	30	P	MQH	TBSD			5×10^{-7}			176
CdTe	35 ~ 80	S	LE	DFWM	6000	580	4.2×10^{-7}		~10	203
CdTe	35	S	LE&CVD	DFWM	~800	580	$\sim6\times10^{-7}$			203
InP	40	S	PP	Z		1064		$\sim7\times10^{-11}$		197
CuCl	30 ~ 160	BS	MQH	TBSD (80 K)			$\sim1\times10^{-6}$		~100	17, 188
CuCl	30 ~ 160	BS	MQH	TBSD (80 K)			$\sim1\times10^{-6}$			115, 188
CuCl	44 ~ 68	BS	MQH	PP (77 K)				$\sim3\times10^{-7}$		182
CuCl	30 ~ 60	S	SG	TBSD (77 K)			1.1×10^{-8}			196
CuCl	~40, ~400	SAS	MQH	TBSD (77 K)		381 ~ 383	$\sim4\times10^{-8}$			199
CuCl	81	BS	MQH	TBSD (77 K)		384	3×10^{-6}			202
CuCl	80	–	MQH	KS (77 K)	94	–	2×10^{-7}		< 4	207
CuBr	182	BS	MQH	TBSD (77 K)		413	$\sim1\times10^{-7}$			202
CuBr	~84	BS	MQH	TBSD (77 K)		–	$\sim2.6\times10^{-7}$		30	190
Cu$_x$S	~80			PP			$\sim10^{-7}$			204
Cu$_x$S				Z			1.3×10^{-9}			205
CuBr$_{0.26}$Cl$_{C.73}$	106 ~ 230	BS	MQH	TBSD (77 K)			1.1×10^{-8}		43	191
PbSe	32 ~ 146	PBS	MQH	Z		1060	$\sim1.14\times10^{-9}$			206

BS, borosilicate; P, phosphate; S, silica, SAS, sodium-almino-silicate; FP, fluoro-phosphate; MQH, melt-quench and heat treatment; PP, porous process; SPT, spattering process; IBS, ion beam sputtering; LE, laser evaporation; SG, sol-gel process; CVD, chemical vapor deposition; PP, pump-probe technique; DFWM, degenerate four wave mixing; KS, Kerr shutter technique; TBSD, two-beam-self-diffraction; FWM, four wave mixing; Z, z-scan.

glass formation temperature to control precipitation of the semiconductor particles [113, 114, 193]. Additionally various methods such as super-cooling, film deposition, porous process, sol-gel process and the ion implantation method have been studied for the development of new semiconductor-doped glass materials. Table 4.6 summarizes the typical nonlinear properties of semiconductor-doped glass materials prepared by various fabrication technologies reported in a large number of papers.

4.3.8 Applications of semiconductor-doped glasses

One example of the practical applications of semiconductor-doped glasses is image reconstruction through optical phase conjugation using commercially available filter glass [14, 126]. The switching characteristic of the nonlinear directional coupler was also demonstrated using waveguide geometry on the CdS_xSe_{1-x}-doped glass made by the ion-exchange process [208]. Bistable switching of the Fabry–Pérot interferometer configuration [129] has been demonstrated using filter glasses as shown in Fig. 4.27. Quantum-confined microcrystallites of PbS in glass were used recently as an intracavity saturable

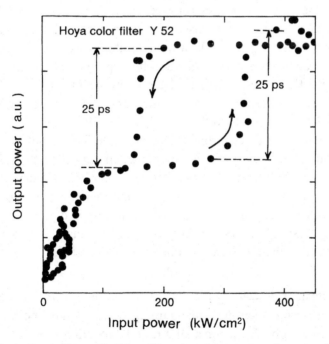

Fig. 4.27. Optical switching with Fabry–Pérot interferometer configuration (CdS_xSe_{1-x} filter glass Y52). [Reprinted from J. Yumoto, S. Fukushima and K. Kubodera, *Opt. Lett.* **12** (1987) 832, copyright (1987) with permission from the Optical Society of America.]

absorber to obtain passive continuous-wave mode locking in a Cr:forsterite laser. Laser pulses with 4.6 ps width were obtained at a 110 MHz repetition rate in this study [209].

4.4 Metal-doped glasses

4.4.1 Fundamentals of optical properties of metal

4.4.1.1 Bulk metal (Drude's theory)

The optical properties of metals are related to the intraband transition of free electrons in the conduction band and interband transitions of electrons from the valence band to the conduction band. Although the intraband transitions have been used to discuss the fundamental absorption bands in semiconductors, the number of free carriers is usually much larger by several orders of magnitude in metals and the absorption related to their intraband transitions is therefore much larger in metals than in semiconductors [210].

The alkali metals are usually regarded as good examples of metals having nearly free electron behavior. They have one valence electron in an s band on each atom: 2s in lithium, 3s in sodium, 4s in potassium. The s band is therefore half-filled and the conduction electrons act as if they were free. The noble metals, such as Cu, Ag and Au are also monovalent metals, but they differ from the alkali metals by having a d shell in the free atoms [102, 211]. As shown in Fig. 4.28, the d band is believed to overlap the s–p conduction band, lying 2.1 to 3.8 eV below Fermi level. In the case of noble metals, therefore, both mechanisms (interband between the d-bands and the conduction band, and intraband transition) contribute to the absorption [210].

If it is assumed that free electrons are responsible for the optical properties of metals, Drude's treatment is more fundamental [36, 210, 212, 213]. This model is based on the assumption that the free electrons of the metal move freely inside the sphere with viscous damping and there are no other forces on the applied electron field. The particle forms a strong dipole in an applied electric field. This situation is described by the following equation of motion, when we assume that the restoring term in Eq. (4.28) is zero,

$$m^*(d^2x/dt^2) + \gamma m^*(dx/dt) = -Ee \qquad (4.64)$$

where γ is a constant describing the viscous damping and m^* is the effective mass of the conduction electrons. The damping arises from the collisions of the electrons and is simply the inverse of the collision time, i.e. $\gamma = 1/\tau$. The steady-state solution of this equation is

$$x = (e/m^*)[E_0 \exp(i\omega t)/(\omega^2 - i\gamma\omega)]. \qquad (4.65)$$

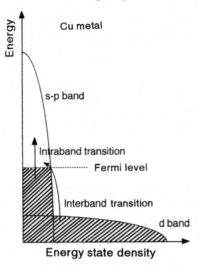

Fig. 4.28. Schematic expression of electron state density and band structure of Cu metal.

The resulting dielectric constant ε_m can then be written as

$$\varepsilon_m(\omega) = \varepsilon'_m - i\varepsilon''_m = 1 - \omega_p^2/\{\omega(\omega - i/\tau)\}. \tag{4.66}$$

Here ω_p denotes the plasma frequency of the bulk material, which is defined by the following equation

$$\omega_p = (4\pi N e^2/m^*)^{1/2} \tag{4.67}$$

depending on the number of conduction electrons per unit volume, N. This means phenomenologically that the optical properties of metals are caused by the oscillation of the electron group density. The equations for the real and imaginary parts of the dielectric constant then become [36, 214, 215]

$$\varepsilon'_m = n^2 - k^2 = 1 - \omega_p^2\tau^2/\{1 + (\omega\tau)^2\} \tag{4.68}$$

$$\varepsilon''_m = 2nk = \omega_p^2\tau/\{\omega(1 + (\omega\tau)^2)\}. \tag{4.69}$$

From the static conductivity of silver or copper, the mean free path of the conduction electron in bulk metal is $\sim$500 Å and $\tau \sim 5 \times 10^{-14}$ s. Therefore, the frequency in the visible and near infrared region is larger than the inverse scattering time $((\omega\tau)^2 \gg 1)$. Then, Eqs. (4.68) and (4.69) are simplified, becoming

$$\varepsilon'_m = 1 - \omega_p^2/\omega^2 \tag{4.70}$$

$$\varepsilon''_m = \omega_p^2/\omega^3\tau. \tag{4.71}$$

Thus, the optical properties of the free carrier are distinguished by the frequency ω relative to ω_p. Usually, the number density N is about 10^{22} cm^{-1}, which yields $\omega_p \sim 10^{16}$ sec^{-1} for noble metals [36, 215]. Therefore, in the case of silver or copper in the visible region, $\omega_p > \omega > 1/\tau$.

4.4.1.2 Small metal particles (mean free path theory) [213, 214, 216, 217]

The optical properties of large particles or bulk metal behave as if they are determined by free electrons, although interband transitions also contribute at these wavelengths. In the case of intermediate size, the incident beam may either be absorbed or scattered. When the diameter d of the sphere is small compared with the wavelength, the scattering losses are negligible. The collisions of the conduction electrons with the particle surfaces become important as an additional process. The new collision relaxation time τ_{eff} due to this process will be given by

$$1/\tau_{eff} = 1/\tau_b + v_F/R \qquad (4.72)$$

where τ_b is the bulk value, v_F is the Fermi velocity of free electrons, and R is the radius of the particles. The Fermi velocity v_F is 1.4×10^6 m/sec for silver and the mean free path $v_F \tau_s$ in bulk material is ~ 500 Å. If $2R \ll \sim 500$ Å, then the mean free path is usually equal to the particle size $2R$ and is entirely limited by the boundary scattering as

$$1/\tau_{eff} = v_F/R. \qquad (4.73)$$

4.4.1.3 Composite materials [213, 214, 216–218]

The optical properties of composite materials, which consist of small spherical metal particles of radius R with a complex dielectric constant ε_m embedded in a dielectric medium such as glass with a real dielectric constant ε_d, are quite different from those of each component material. The structure of the composite material is shown schematically in Fig. 4.29.

When the diameter of the metal spheres is small compared with the wavelength and the volume fraction p occupied by the metal particles is small compared with 1, ($p \ll 1$), the polarization of such a composite material may be described in terms of the effective dielectric constant ε_{eff}, which is written as the approximate form [213, 218] originally derived by J. C. Maxwell-Garnett [219, 220]

$$\varepsilon_{eff} = \varepsilon_d + 3p\varepsilon_d \{(\varepsilon_m - \varepsilon_d)/(\varepsilon_m + 2\varepsilon_d)\} \qquad (4.74)$$

where $\varepsilon_m = \varepsilon'_m + i\varepsilon''_m$ (complex and frequency dependent) is the dielectric constant of the metallic particle and ε_d (real) is that of the host dielectric medium. The absorption coefficient of the composite materials related to the imaginary part of ε_{eff} is easily deduced [214, 216, 218] from Eq. (4.74)

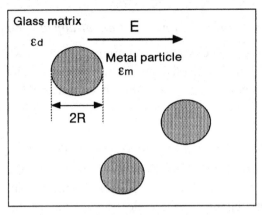

Fig. 4.29. Schematic diagram of a composite material consisting of a metal particles of radius R with complex dielectric constant ε_m embedded in a dielectric matrix with a real dielectric constant ε_d.

$$\alpha = (2\omega/c)k$$
$$= (9 p\omega\varepsilon_d^{3/2}/c)[\varepsilon_m''/\{(\varepsilon_m' + 2\varepsilon_d)^2 + \varepsilon_m''^2)\}]. \qquad (4.75)$$

The absorption coefficient has a maximum value at the frequency ω_s for which

$$\varepsilon_m'(\omega_s) + 2\varepsilon_d + 0. \qquad (4.76)$$

This is called 'surface plasma resonance', corresponding to the collective electronic excitations at the interface between a metal and a dielectric. As in the vicinity of ω_s, from Eqs. (4.70), (4.71), (4.75) and (4.76), the absorption coefficient can be expressed [216]

$$\alpha = \{9 p\varepsilon_d^{3/2}/4c(2\varepsilon_d + 1)^3\}[(\omega_p^2/\omega^2\tau)/\{(\omega_s - \omega)^2 + (2\tau)^{-2}\}] \qquad (4.77)$$

The absorption spectrum has a Lorentzian shape and the full width γ at half-height of this spectrum is just the inverse of the scattering time τ. When the diameter of the metal particles is sufficiently small, the scattering time is given by Eq. (4.73). Consequently one might expect that the width of the surface plasma resonance γ should be proportional to $1/R$, leading to a broadening of the surface plasma width for smaller particles. For instance, gold and copper particles typically in the range 10 to 100 nm in diameter are ruby red with an absorption band peaking at $\sim$550 nm [221, 222]. They also have absorption due to the interband transition overlapping the plasma resonance peak. The beautiful colors of such composite materials have been used in the art of colored stained glass for churches or cathedrals and red colored goblets.

4.4.2 Optical nonlinear mechanisms of metal-doped composite materials

The optical response of composite materials which consist of small metal particles embedded in a dielectric medium is mainly affected either by an electric-field enhancement effect (dielectric confinement) or by a quantum confinement effect.

4.4.2.1 Local electric field effect (dielectric confinement) [218, 223–225]

The third-order nonlinear polarization $P^{(3)}(\omega)$ induced by a strong optical field E is characterized by a change of the effective dielectric constant $\delta\varepsilon_{eff}$. If this change results from a change of the dielectric constant of the metallic particles $\delta\varepsilon_m$, the relation between the two is obtained by differentiating Eq. (4.74),

$$\delta\varepsilon_{eff} = \{3\varepsilon_d/(\varepsilon_m + 2\varepsilon_d)\}^2 p\delta\varepsilon_m = f_2 p\delta\varepsilon_m. \tag{4.78}$$

If $\delta\varepsilon_m$ is only due to the $\chi^{(3)}$ process of the electrons in the metal spheres, a change $\delta\varepsilon_m$ is simply related to the third-order susceptibility $\chi_m^{(3)}$ of the metal by the relation

$$\delta\varepsilon_m = 12\pi\chi_m^{(3)} E_{in}^2 \tag{4.79}$$

where E_{in} is the local field inside the metallic particles. When a metal particle is placed in a uniform electric field E, the field inside the particle is derived by local field correction theory,

$$E_{in} = \{3\varepsilon_d/(\varepsilon_m + 2\varepsilon_d)\} E = f_i E. \tag{4.80}$$

Here f_i is the ratio between the internal field E_{in} and the external field E, and referred to as the local field factor. Since $f_2 = f_i^2$, Eq. (4.78) can be written as

$$\delta\varepsilon_{eff} = 12\pi p f_i^4 \chi_m^{(3)} E^2. \tag{4.81}$$

The third-order nonlinear polarization thus can be written as

$$P^{(3)}(\omega) = 3p f_i^4 \chi_m^{(3)} E^3 = \chi^{(3)} E^3 \tag{4.82}$$

$$\chi^{(3)} = 3p|f_i|^4 \chi_m^{(3)}. \tag{4.83}$$

The local field factor is generally smaller than 1, but at the surface plasma resonance frequency ω_s for which $\varepsilon_m'(\omega_s) + 2\varepsilon_d = 0$, a dramatic enhancement of the local field inside the particle can be achieved.

In the plasma resonance condition, the focal field factor becomes

$$f_i(\omega) = 3\varepsilon_d/|\varepsilon_m''(\omega)|. \tag{4.84}$$

From Eqs. (4.71) and (4.72), on the other hand, the size dependence of ε_m'' is given by [214]

$$\varepsilon_m''(\omega) = \varepsilon_b''(\omega) + A(\omega)/R \tag{4.85}$$

where $\varepsilon_b''(\omega)$ is the imaginary part of the dielectric constant of bulk metal, R is the particle radius and $A(\omega)$ is a frequency-dependent parameter. Then if we assume that $\chi_m^{(3)}$ is independent of particle size, the third-order nonlinear susceptibility is expected to increase as the volume fraction of the metal particle and the mean particle size increase.

4.4.2.2 Quantum mechanical effect (quantum confinement) [224–227]

Another essential contribution to the nonlinear susceptibility is $\chi_m^{(3)}$. Three quantum mechanisms have been proposed to make major contributions to $\chi_m^{(3)}$ for noble metal particles: intraband transition, interband transition and hot electron [224–227].

A first contribution to $\chi_m^{(3)}$ arising from the conduction electrons is the contribution of intraband transitions between filled and empty states in the conduction band, corresponding to the Drude dielectric constant. Due to size quantization, the conduction band will be split into discrete levels with an average separation that is large compared with thermal energies. This effect is called quantum size effect and has already been discussed in the section on semiconductor-doped glasses. When the radius R of the metal sphere is in the range $5 \sim 50$ nm the component of $\chi_m^{(3)}$ due to this new contribution comes into play. It is approximately given by

$$\{\chi_m^{(3)}{}_{\text{intra}}\} \sim -\mathrm{i}(1/R^3)\{e^4 E_F^4/m^2\hbar^5\omega^7\}g(v)(1 - d/d_0) \qquad (4.86)$$

where e and m are the charge and mass of a free electron, E_F is the Fermi energy, $g(v)$ is a kind of shape factor and usually a number of order 1, the parameter d_0 is the diameter representing the upper limit of the quantum size effect. The value of d_0 is given approximately by $d_0 \sim T_2(2E_F/m)^{1/2}$, where T_2 is the dephasing time [224]. In the case of Cu, this leads to a value of d_0 around 30 nm [226], comparable with that for Au [224]. For much smaller R $(d < d_0)$, $\chi_m^{(3)}$ is thus size-dependent on R^{-3}. The value of the contribution of the conduction electron to the nonlinear susceptibility was expected to be of the order of 10^{-13} esu [224].

In addition to the free electron intraband term, the dielectric constant of a noble metal contains a term corresponding to the interband transitions between the d and s–p bands. A contribution to $\chi_m^{(3)}$ arising from this interband transition will consist of the fully resonant contribution of the two corresponding levels. In other words, it corresponds to the saturation of the two-level transition. The imaginary part of $\chi_m^{(3)}$ responsible for interband transition has the form [224]

$$\mathrm{Im}\{\chi_m^{(3)}{}_{\text{inter}}\} \sim -(e^4/5\hbar^2 m^4\omega^4)T_1' T_2'|r_{\text{dc}}|^4 \qquad (4.87)$$

where T_1' and T_2' are the energy lifetime and the dephasing time for a two-level system, r_{dc} is the matrix element of transition. From this equation, $\chi_m^{(3)}{}_{\text{inter}}$ is slightly size-dependent because the dephasing time T_2' changes with the dimensions of the metal particles. The d-band electrons have very large effective masses and they are already localized. The d-band electrons are therefore not strongly affected by the quantum confinement but only affected for very small spheres with sizes < 2.5 nm. Assuming the values of each parameter, the contribution from this term was estimated to be of the order of 10^{-8} esu [224].

The third possible mechanism contributing to $\chi_m^{(3)}$ is the nonequilibrium electron heating. When a laser beam whose frequency ω is close to the surface plasma resonance frequency, ω_s, is incident upon a metal particle, interband or intraband transition results in high electronic temperatures, leading to a deformation of the Fermi-Dirac distribution of electrons near the Fermi surface, i.e. it increases the electron occupancy above the Fermi energy and decreases the occupancy below. The so-called Fermi smearing affects the transition probability of the d-band electrons to the conduction-band energies near the Fermi level, which changes the dielectric function and modifies the absorption coefficient of d-electron. The Fermi-smearing also results in an enhanced probability for scattering between states in the vicinity of the Fermi energy. This enhancement of the total electronic scattering rate leads to a surface plasma line broadening due to surface-scattering-induced damping.

The Fermi smearing or hot-electron contribution to $\chi_m^{(3)}$ is given by [224]

$$\mathrm{Im}\{\chi_m^{(3)}{}_{\mathrm{hot}}\} \sim (1/24\pi^2\gamma T)(\omega\varepsilon_D''\tau_0)(\partial\varepsilon_L''/\partial T) \tag{4.88}$$

where γT is the specific heat of the conduction electron, ε_D'', and ε_L'' is the imaginary part of the dielectric constant due to the free electron and interband transition, respectively, τ_0 is the time constant for the cooling of electrons. This so-called hot-electron effect is a thermal effect. Therefore the changes in $\chi_m^{(3)}$ induced by this effect are expected to be size independent. The predicted value of $\chi_m^{(3)}$ due to Fermi smearing is expected to be of the order of $\sim 10^{-7}$ esu [224].

4.4.3 Nonlinear optical properties of metal-doped glasses

4.4.3.1 Nonlinear susceptibility, $\chi^{(3)}$

As discussed in the previous section, the large third-order susceptibility of these glasses is predominantly attributed to local field enhancement near the surface plasma resonance of the metal particles [223–225]. Figure 4.30 shows a typical example of transmission electron micrographs of Cu-doped glass. Most of the metal particles appear as dark spheres roughly 10 nm in diameter. Figure 4.31 shows representative absorption spectra and the wavelength dependence of $\chi^{(3)}$ of Cu- and Ag-doped glasses [228]. A broad plasma resonance peak and, in addition, a large absorption band are observed, for

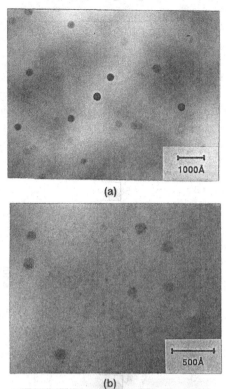

(a)

(b)

Fig. 4.30. Transmission electron micrograph of Cu- and Ag-doped BaO–P$_2$O$_5$ glass. (a) 50P$_2$O$_5$–50BaO–6SnO–6Cu$_2$O, (b) 50P$_2$O$_5$–50BaO–4SnO–4Ag$_2$O. [Reprinted from K. Uchida, S. Kaneko, S. Omi, C. Hata, H. Tanji, Y. Asahara, *et al. J. Opt. Soc. Am.* **B11** (1994) 1236, copyright (1994) with permission from the Optical Society of America.]

instance, in the copper-doped glasses. The values of $\chi^{(3)}$ denoted by filled circles exhibit a peak at almost the same wavelength as the absorption peak. The maximum value of $|\chi^{(3)}|$ is of the order of 10^{-7} esu for both types of metal particles.

4.4.3.2 *Nonlinear susceptibility of metal particle,* $\chi_m^{(3)}$

By using the local field factor calculated from Eq. (4.80) and the volume fraction p extracted from expression (4.75) for the absorption coefficient, the value of $\chi_m^{(3)}$ can be calculated from the $\chi^{(3)}$ data through Eq. (4.83). The results were, for example, $(2 \sim 4) \times 10^{-9}$ esu for silver [223, 228], $\sim 10^{-6}$ for copper [228] and $\sim 8.0 \times 10^{-9}$ esu for gold particles [223, 224]. The $\chi_m^{(3)}$ for copper particles is much larger than that for silver and gold particles, which may be due to the difference in the hot electron and interband contributions to the

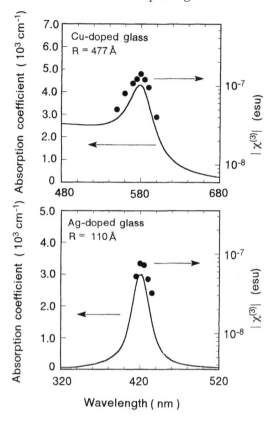

Fig. 4.31.   Absorption spectra and wavelength dependence of $\chi^{(3)}$ for Cu- and Ag-doped glasses. [Reprinted from K. Uchida, S. Kaneko, S. Omi, C. Hata, H. Tanji, Y. Asahara, *et al. J. Opt. Soc. Am.* **B11** (1994) 1236, copyright (1994) with permission from the Optical Society of America.]

plasma resonance. Copper has a high density of states correlated with the d-band, which enhance the hot-electron contribution owing to Fermi smearing at the plasmon frequency.

The estimated value of $\chi_m^{(3)}$ in the Au-doped glasses prepared by the sol-gel process was exceptionally in the range of $10^{-7} \sim 10^{-6}$ esu [229–231]. These values are about one to three orders of magnitude larger than the reported values of $\sim 10^{-9}$ esu in Au-doped glasses prepared by melting and subsequent heat treatment. Such large values of $\chi_m^{(3)}$ for these glasses are now thought to be due to the nonequilibrium state of the Au particles, but the errors from uncertainties of dielectric model, size distribution, etc. in calculation of $\chi_m^{(3)}$ may also have contributed [232].

Size dependence of $\chi_m^{(3)}$ has also been studied by many workers. The value of $\chi_m^{(3)}$ for copper ($R = 2.5 \sim 47.7$ nm) [228], for silver ($R = 2.1 \sim 15.3$ nm) [228] and for gold ($d = 2.8 \sim 30$ nm [224] and $d = 5 \sim 30$ nm [233]) particles determined experimentally were roughly independent of particle size, which is consistent with the Fermi smearing or hot electron contribution to $\chi_m^{(3)}$. In contrast, Li Yang et al. [226] reported recently in their experimental data that $\chi_m^{(3)}$ of the Cu clusters in the Cu-implanted SiO$_2$ is inversely proportional to d^3 for $d < 10$ nm, as shown in Fig. 4.32. Therefore, when the diameters of the Cu particles are smaller than ~ 10 nm, the intraband transition in the conduction band probably makes the main contribution to $\chi_m^{(3)}$.

4.4.3.3 Response time

If the nonlinear response of metal particles is based on nonequilibrium electron heating, the response time is determined by the cooling time of the hot electron, which cools to the metal lattice through electron–phonon scattering in $2 \sim 3$ ps [234]. In fact most of the available literature on silver and gold colloids have reported that the nonlinear response is fast on a several-ps time scale [223, 225]. The time response of the nonlinear conjugate signal for the gold-doped glass consists of two components. One component is faster than the laser pulse duration of 30 ps, which is likely due to the cooling process of nonequilibrium heating of the electron gas (Fermi smearing) to the gold lattice. The slower component is thought to be due to residual heating of the gold

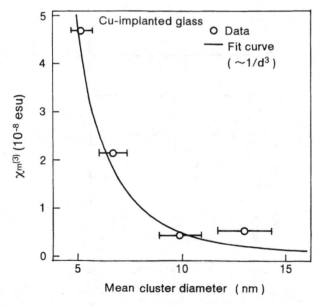

Fig. 4.32. $\chi_m^{(3)}$ of the Cu clusters as a function of mean particle diameter for Cu-implanted SiO$_2$. [Reprinted from Li Yang, K. Becker, F. M. Smith, R. H. Magruder III, R. F. Haglund, Jr., Lina Yang, et al. J. Opt. Soc. Am. **B11** (1994) 457, Copyright (1994) with permission from the Optical Society of America.]

lattice. The cooling time of the gold lattice increases with particle size and 115 ps is predicted for a 30 nm-sized gold particle. A similar nonlinear response consisting of fast and slow components was also found in experiments using Cu-doped glasses [228, 235].

More recently pump-probe experiments with a time resolution of 100 fs have been performed for copper particles with a radius of 4 nm [236]. Figure 4.33 (a) shows the linear absorption spectrum of copper particles with a radius of 4 nm at 300 K. A surface plasma peak is observed at 2.2 eV. A large shoulder is due to the transition from the d band. Figure 4.32 (b) shows differential absorption spectra for various delay times. When the lower energy side of the plasma peak is excited by the pumping pulses, absorption bleaching is observed at the plasma peak and the increase emerges at both sides of the peak. This spectral behaviour is just the broadening of the plasma absorption band, suggesting that the hot electrons are the dominant contribution to the measured nonlinearity.

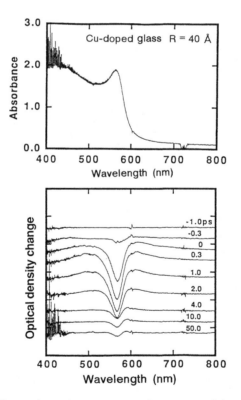

Fig. 4.33. (a) The linear absorption spectrum of copper particles with a radius of 4 nm at 300 K. (b) The differential absorption spectra for various delay times. [Reprinted from T. Tokizaki, A. Nakamura, S. Kaneko, K. Uchida, S. Omi, H. Tanji *et al. Appl. Phys. Lett.* **65** (1994) 941, copyright (1994) with permission from the American Institute of Physics.]

Figure 4.34 shows the nonlinear response time at the early stage of the time delay for different laser fluences derived from the recovery time of the absorption spectra [236]. The decay behavior of the bleaching signal exhibits nonexponential decay with two components. The fast component is decayed within 5 ps and the slow one remains for times of more than 150 ps. The fast one depends on the pumping laser fluences, and is as short as 0.7 ps for 0.21 mJ/cm^2. These behaviors are explained as the relaxation dynamics of electron temperature and lattice-effective temperature by the usual electron-photon coupling model.

A similar pronounced broadening of the surface plasma line upon interband excitation and direct excitation of the surface plasmon has also been studied in gold colloids embedded in a sol-gel matrix [237] and Ag colloids embedded in a glass [238–240]. The recovery dynamic was characterized by a two-component decay with time constants of 4 and ~200 ps [237]. It was shown in the gold colloids that pumping at interband transition frequencies results in a more efficient surface plasma damping than at the surface plasma resonance pumping itself [237]. In the Ag colloid case, however, the direct excitation of the surface plasma was more efficient [238–240].

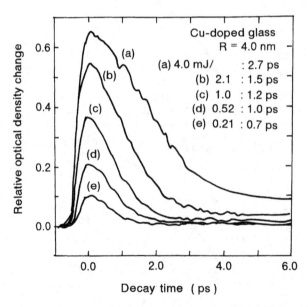

Fig. 4.34. Time response at the early stage of the time delay for various laser fluences. [Reprinted from T. Tokizaki, A. Nakamura, S. Kaneko, K. Uchida, S. Omi, H. Tanji *et al. Appl. Phys. Lett.* **65** (1994) 941, copyright (1994) with permission from the American Institute of Physics.]

4.4.3.4 Figure of merit

The nonlinear material is evaluated in terms of large $\chi^{(3)}$ which is produced by the large optical absorption, whereas a high throughput is also required in the practical nonlinear devices. This leads to a materials figure of merit for nonlinear devices defined as

$$F = \chi^{(3)}/\alpha. \tag{4.89}$$

Near the plasma resonance frequency ω_s, if $\varepsilon_d \sim n_g^2$, the local field factor given by Eq. (4.84) becomes

$$f_i(\omega_s) = 3\varepsilon_d/\varepsilon_m'' = 3n_g^2/\varepsilon_m'' \tag{4.90}$$

where n_g is the refractive index of the glass matrix. The absorption coefficient, Eq. (4.75) is given by [224]

$$\alpha = p(\omega/nc)|f_i|^2\varepsilon_m''. \tag{4.91}$$

The materials figure of merit, $\chi^{(3)}/\alpha$, can now be written as

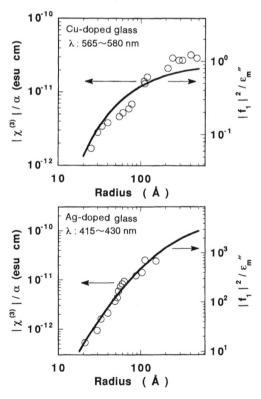

Fig. 4.35. $\chi^{(3)}/\alpha$ and f_i^2/ε_m'' as a function of particle size. [Reprinted from K. Uchida, S. Kaneko, S. Omi, C. Hata, H. Tanji, Y. Asahara, *et al. J. Opt. Soc. Am.* **B11** (1994) 1236, copyright (1994) with permission from the Optical Society of America.]

$$\chi^{(3)}/\alpha = (cn/\omega)|f_i|^2|\chi_m^{(3)}|/\varepsilon_m''(\omega)$$
$$= 9n_g^5(c/\omega_s)|\chi_m^{(3)}|/|\varepsilon_m''(\omega)|^3. \qquad (4.92)$$

Since the size dependence of ε_m'' is given by Eq. (4.85), a large figure of merit is expected by increasing the particle radius or by increasing the refractive index of the glass material.

Figure 4.35 shows a typical example of the figure of merit $\chi^{(3)}/\alpha$ of glasses that have been doped with copper and silver particles measured at the absorption peak as a function of particle radii [228]. The value of $\chi^{(3)}/\alpha$ increases with an increase of radius for the copper particles ($R < 30$ nm) and for silver particles ($R < 15$ nm). The solid curves in the figure represent the calculation of f_i^2/ε_m'' based on the mean free path theory. Excellent agreement is obtained between the slope of the measured $\chi^{(3)}/\alpha$ and the calculated f_i^2/ε_m'' as expected from Eqs. (4.90) and (4.92). Such an enlargement of $\chi^{(3)}/\alpha$ with an increase of particle radius was also reported for Au-doped SiO_2 film prepared by a sputtering method [241–243].

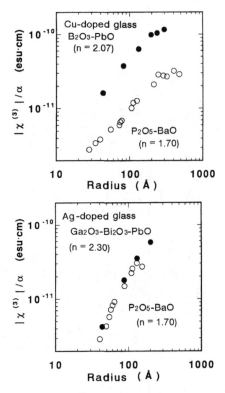

Fig. 4.36. The measured value of $\chi^{(3)}/\alpha$ as a function of particle size for various refractive indices of matrix glasses.

Table 4.7. *Nonlinear characteristics of metal-doped glasses made by new fabrication processes*

Dopant	Matrix	Diam. (nm)	Process	Measurement (λ, nm)	α (cm^{-1})(λ, nm)	$|\chi^{(3)}|$ (esu)	τ (ps)	$\chi_m^{(3)}$ (esu)	Ref.
Au	glass	2.8 $\sim$ 30	MQH	DFWM	($\sim$530)	$\sim$5.5 (10^{-11})	5	$\sim$8.2 (10^{-8})	224
		10	MQH	DFWM (532)			28	1.5 (10^{-8})	223
	SiO$_2$	5.5	SPT	DFWM (532)					246
	SiO$_2$	5.8	IP	DFWM (532)		1.3 (10^{-7})			247
	SiO$_2$	$\sim$3	IP	DFWM (532)		1.2 (10^{-7})			252
	SiO$_2$	5.8	SPT	DFWM (532)		$10^{-7} \sim 10^{-8}$		8 (10^{-8})	249
	SiO$_2$	8.8	MQH	DFWM (532)		3.5 (10^{-8})		2.5 (10^{-8})	249
	SiO$_2$	1.4 $\sim$ 6	SG	DFWM (532)	3 (10^3) (450)	2.5 (10^{-11})		3 (10^{-8})	229
	SiO$_2$	5 $\sim$ 30	IP	DFWM (532)		$\sim$7.7 (10^{-9})		7 (10^{-8})	248
	SiO$_2$		SG	TBSD	$\sim$5.5 (10^5) (450)	1.7 (10^{-10})		10^{-7}	230
	Water	5 $\sim$ 30	SPT	DFWM (532)		$\sim$6.3 (10^{-8})			233
	SiO$_2$		SG	DFWM (532)			29	$\sim$2 (10^{-6})	242
	SiO$_2$	10	SPT	TBSD (540)	2.6 (10^3)			2 $\sim$ 5 (10^{-8})	231
	SiO$_2$	3 $\sim$ 80	SPT	DFWM (532)		2.3 (10^{-8})		5 $\sim$ 20 (10^{-8})	259
	SiO$_2$	3 $\sim$ 33.7	SPT	DFWM (530)		2.5 (10^{-6})		1.1 (10^{-6})	243
	SiO$_2$		SPT	DFWM (532)		2.0 (10^{-7})		3.0 (10^{-7})	260
Ag	glass	7	SPT	DFWM (400)		$\sim$4 (10^{-6})		$\sim$4.2 (10^{-8})	223
	glass	4 $\sim$ 30	MQH	TBSD (532)	$\sim$3.5 (10^3) (532)	$\sim$7.6 (10^{-8})		2.4 (10^{-9})	228
	glass		IE, IP	ZS (591)		$n_2'' = (10^{-10})(\text{cm}^2/\text{W})$		2 $\sim$ 4 (10^{-9})	258

Metal	Matrix	Size (nm)	Preparation	Measurement (nm)				Ref.
Cu	SiO_2	2 ~ 5	SPT	TBSD (400)	5.7 (10^4) (400)	1.6 (10^{-8})		256
	glass		MQH	TBSD			0.75 –	264
	glass	5 ~ 100	MQH	TBSD (532)	~4.7 (10^3) (532)	1.2 (10^{-7})	1 ~ 2.5 (10^{-6})	228
	glass	8	MQH	PP		3.5 (10^{-9}) 0.7	1.4 (10^{-6})	236
	SiO_2	2 ~ 28	IP	ZS (532)		$n_2'' = {\sim}2.3\ (10^{-7})(cm^2/W)$	<5	235
	SiO_2		IP	ZS (570 ~ 600)	~(10^4)	~(10^{-8})		250
	SiO_2	5.2 ~ 12.6	IP	DFWM (532)		$2.0 < n_2'' < 4.2\ (10^{-10})\ (cm^2/W)$		226
	SiO_2	12 ~ 28	SPT	DFWM (532)		~8.4 (10^{-10})	~46.8 (10^{-9})	251
	SiO_2	5.2 ~ 12.6	IP	ZS (570)		4 (10^{-9})		259
	SiO_2	~10	IP	ZS (770)	1.2 (10^4)	$n_2'' = {\sim}6\ (10^{-10})\ (cm^2/W)$		262
Au–Cu	SiO_2	~30	IP	ZS (570)	4.6 ~ 5.9 (10^4)	$n_2'' = 5\ (10^{-11})\ (cm^2/W)$		254
Ag–Cu	SiO_2	1 ~ 10	SG	DFWM (532)		$n_2'' = {\sim}1.6\ (10^{-9})(cm^2/W)$		261
P	SiO_2	3 ~ 5	IP	TBSD (390)		~1.06 (10^{-7})		253
Sn	SiO_2	4 ~ 20	IP	TBSD		1.2 (10^{-6})		255
Ni	SiO_2	~10	IP	ZS (770)	1.4 (10^4)	3 (10^{-6}) $n_2'' = 1.7\ (10^{-10})\ (cm^2/W)$		262
Cu–Ni	SiO_2	~10	IP	ZS (770)	4.0 (10^4)	$n_2'' = 6.8\ (10^{-10})\ (cm^2/W)$		262
Co	SiO_2		IP	ZS (770)		$n_2'' = 1.8\ (10^{-10})\ (cm^2/W)$		263

MQH, melt-quench and heat-treatment; SPT, spattering; IP, ion implantation; IE, ion exchange; SG, sol-gel process; PP, Porous process.
DFWM, degenerate four-wave mixing; TBSD, two beam self diffraction; ZS, z-scan technique; PP, pump-probe.

The influences of the refractive index of the matrix materials on the figure of merit were also investigated, according to Eq. (4.92). Figure 4.36 shows the figure of merit of Cu- and Ag-doped glasses as a function of metal particle size for matrix glasses with various refractive indices [244]. For the Cu-doped glass, it is seen that the figure of merit increases 2.07 times as an increase of refractive index from 1.7 to 2.07. The same effect of the enlargement of the figure of merit was also found for Au-doped composite materials [241, 242]. In contrast, such enhancement in the figure of merit cannot be observed for Ag-doped glass [244].

According to Eq. (4.92), $\chi^{(3)}/\alpha$ is simply proportional to the n_g, inverse of ω_s and ε_m''. The figure of merit, therefore, increases also with a decrease of ω_s accompanying the increase of n_g. This factor is only 1.04 for Cu particles and 1.21 for Ag particles, and negligible in both cases. The value of ε_m'' also depends on the frequency and decreases with the shift of the plasma resonance frequency ω_s [245]. For the Ag-doped glass, the figure of merit increases 4.5 times when the refractive index of the matrix increases from 1.7 to 2.3. The value of ε_m'', in contrast, increases with the shift of resonant frequency ω_s. The factor induced by this change of ε_m'' is 0.21, which compensates the increase due to the refractive index changes [244].

4.4.4 Fabrication techniques

Glasses containing small metal particles have been prepared by various methods such as melt-quenching, the sol-gel process, sputtering and the ion implantation process. Table 4.7 summarizes the main nonlinear properties of glass materials prepared by various fabrication techniques.

4.5 Comparisons of optical nonlinearities of various nonlinear glass materials

As stated in Section 4.1.2, the optical nonlinearities of the glass materials can be divided into two principal categories, resonant or nonresonant, depending on whether the transition generates real or virtual excited-state populations. Table 4.8 summarizes the classification and main nonlinear mechanisms of some representative nonlinear glass materials. Figure 4.37 summarizes and compares the nonlinear susceptibilities and nonlinear response times of various nonlinear glass materials. Figure 4.38 compares the various nonlinear glass materials in terms of the optical excitation process.

The dominant intrinsic or nonresonant nonlinearities in all glass materials are produced by excitation in the transparency region well below the band gap.

Table 4.8. *Classification and main nonlinear mechanisms of representative nonlinear glass materials*

Dopant	Carrier	Mechanism of optical nonlinearity	
		Dopant	Glass materials – Size effect
Nonresonant • High index glasses • Chalcogenide glasses Resonant • CuCl and CuBr-doped glasses	Insulator • Bound electron	• Unharmonic terms of electronic polarization	
• CuCl and CuBr-doped glasses	Wide-gap semiconductor (Bound exciton)	• Biexciton • Phase space filling • Plasma screening • Exchange effect	Weak quantum confinement • Enhancement of space filling • Super-radiation • Giant oscillation
• CdSSe, CdSe and CdTe-doped glasses • Color filter glasses	Middle-gap semiconductor (Electron and hole)	• Band filling effect (Blocking effect) • Auger recombination • Free electron absorption • Trapping effect • Thermal effect	Strong quantum confinement • Enhancement of state filling • Auger recombination • Free electron absorption • Trapping effect • Photodarkening effect
• Ag, Cu, and Au-doped glasses	Conductor (Free electron)	• Hot electron (Fermi smearing) • Saturation of interband and intraband absorption	Dielectric confinement • Local field effect • Hot electron scattering (Fermi smearing) Quantum confinement • Enhancement of saturation of intra- and inter-band absorption

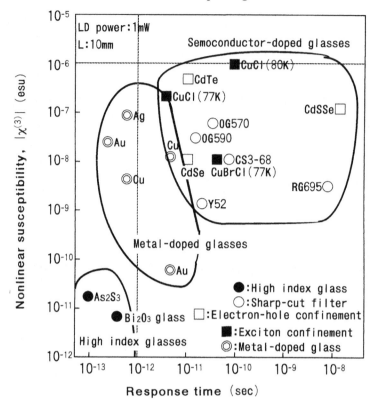

Fig. 4.37.   The nonlinear susceptibilities versus nonlinear response time. (Data from Table 4.4,
4.5, 4.6 and 4.7. The broken lines (response time: 10^{-12} sec, $|\chi^{(3)}|$: 10^{-6} esu) indicate the
conditions required for obtaining 2π phase shift when pulse selector is made using 1 mW-LD and
10 mm optical path [265].)

They have advantages of negligible linear loss in the wavelength range of
interest. The nonlinearities are due to anharmonic motion of bound electrons
induced by the optical fields. The driven components are due to 'virtual
populations' as long as the field is applied. In this process, there is no energy
deposition in the material and no participation in any relaxation process. They
are characterized by an extremely fast response by a simple application of the
uncertainty principle, but the field-induced optical nonlinearities are usually
small.

The remaining nonlinearities are resonant, which means electronic movement from one energy level to another. The real excitations generate predominantly large excited-state populations of the free carriers for a given excitation
intensity and modify the absorption properties. The absorption change $\Delta\alpha(\omega')$
gives the refractive index changes $\Delta n(\omega)$ through the K–K transformation (see
Eq. (4.16)) which lead to significant third-order optical nonlinear effects.

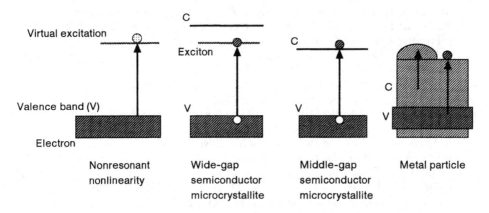

Fig. 4.38. Various nonlinear optical processes in nonlinear glasses.

Glasses dispersed with semiconductor microcrystallites or metal microparticles fall into this category.

In the semiconductor-doped glasses, the band-filling effect or blocking effect mainly produces the nonlinear absorption changes leading to the optical nonlinearities. The quantum size effect in the microcrystallites, i.e. the state-filling effect leads to a further enhancement of these nonlinearities (state filling). A major advantage of these glasses is the large nonlinear effects with only a small optical field. However, since the populations have a finite lifetime and participate in a number of relaxation processes, response times are consequently limited by speeds corresponding to energy build up and relaxation times. Carrier trapping, Auger recombination, free carrier absorption and photodarkening effects additionally affect the nonlinear response time.

For the specific case of semiconductor-doped glasses (wide-gap semiconductor), a bound state of the electron-hole pair (exciton state) analogous to the nonresonant electric polarization is conserved in the microcrystallites, which results in much higher nonlinearities and a fast nonlinear response. This situation, however, can only be achieved at low temperature.

The last candidate is glasses doped with small metal particles, such as Cu, Ag, and Au. In this case, the nonlinearities are mainly attributed to local field enhancement near the surface plasma resonance and nonequilibrium electron heating (hot electron). Since the effect is caused by the collective motion of free electrons, they exhibit large optical nonlinearities as well as fast response times near the surface plasma resonance frequency.

References

1 E. M. Vogel, M. J. Weber and D. M. Krol: *Phys. Chem. Glass.* **32** (1991) 231.
2 R. K. Jain and M. B. Klein: Chap. 10 in *Optical Phase Conjugation*, ed. R. A. Fisher, (Academic Press Inc., San Diego, 1983) p. 307.
3 A. J. Dekker: Chap. 6 in *Solid State Physics*, (Prentice Hall, Inc., Englewood Cliffs and Maruzen Company, Ltd., Tokyo, 1957) p. 133.
4 A. Yariv and R. A. Fisher: Chap. 1 in *Optical Phase Conjugation*, ed. R. A. Fisher, (Academic Press, Inc., San Diego, 1983) p. 1.
5 P. N. Butcher and D. Cotter: Chap. 1 in *The Elements of Nonlinear Optics*, (Cambridge University Press, Cambridge, 1990) p. 1.
6 Y. R. Shen: Chap. 4 in *The Principles of Nonlinear Optics*, (John Wiley and Sons, Inc., New York, 1984) p. 53.
7 S. Singh: *CRC Handbook of Laser Science and Technology*, ed. M. J. Weber, Vol. III Optical Materials Part 1, Nonlinear Optical Properties/Radiation Damage (CRC Press, Inc., Boca Raton, Florida, 1986) p. 3.
8 W. L. Smith: *CRC Handbook of Laser Science and Technology*, ed. M. J. Weber, Vol. III Optical Materials Part 1, Nonlinear Optical Properties, (CRC Press, Inc., Boca Raton, Florida, 1986) p. 259.
9 W. L. Smith: ibid, p. 229.
10 D. A. B. Miller, C. T. Seaton, M. E. Prise and S. D. Smith: *Phys. Rev. Lett.* **47** (1981) 197.
11 G. R. Olbright and N. Peyghambarian: *Appl. Phys. Lett.* **48** (1986) 1184.
12 G. R. Olbright, N. Peyghambarian, S. W. Kock and L. Banyai: *Opt. Lett.* **12** (1987) 413.
13 R. K. Jain and M. B. Klein: *Appl. Phys. Lett.* **35** (1979) 454.
14 R. K. Jain and R. C. Lind: *J. Opt. Soc. Am.* **73** (1983) 647.
15 J. J. Wynne: *Phys. Rev.* **178** (1969) 1295.
16 M. A. Khan, P. W. Kruse and J. F. Ready: *Opt. Lett.* **5** (1980) 261.
17 A. Nakamura, T. Tokizaki, H. Akiyama and T. Kataoka: *J. Lumin.* **53** (1992) 105.
18 N. Finlayson, W. C. Banyai, C. T. Seaton, G. I. Stegeman, M. O'Neill, T. J. Cullen *et al.*: *J. Opt. Soc. Am.* **B6** (1989) 675.
19 P. D. Maker, R. W. Terhune, M. Nisenoff and C. M. Savage: *Phys. Rev. Lett.* **8** (1962) 21.
20 R. Loudon: *The Quantum Theory of Light*, (Oxford University Press, Oxford, 1973) p. 321.
21 K. Kubodera and H. Kobayashi: *Mol. Cryst. Liq. Cryst.* **182A** (1991) 103.
22 K. Kubodera and T. Kaino: *Nonlinear Optics of Organics and Semiconductors*, ed. T. Kobayashi (Springer-Verlag, Berlin, 1989) p. 163.
23 M. Sheik-Bahae, A. A. Said and E. W. Van Stryland: *Opt. Lett.* **14** (1989) 955.
24 M. Sheik-Bahae, A. A. Said, T. H. Wei, D. J. Hagan and E. W. Van Stryland: *IEEE J. Quantum Electron.* **26** (1990) 760.
25 G. I. Stegeman, E. M. Wright, N. Finlayson, R. Zanoni and C. T. Seaton: *J. Lightwave Technol.* **6** (1988) 953 and refs cited therein.
26 A. Yariv: Chap. 19 in *Quantum Electronics*, Third edn, (John Wiley and Sons Inc., New York, 1989) p. 495.
27 Y. R. Shen: Chapter 14 in *The Principles of Nonlinear Optics*, (John Wiley and Sons, Inc., New York, 1984) p. 242 and refs. cited therein.
28 A. E. Siegman, P. A. Belanger and A. Hardy: Chap. 13 in *Optical Phase Conjugation*, ed. R. A. Fisher, (Academic Press Inc, San Diego, 1983) p. 465.

29 T. R. O'Meara, D. M. Pepper and J. O. White: Chap. 14 in ibid, p. 537.

30 S. R. Friberg, Y. Siberberg, M. K. Oliver, M. J. Andrejco, M. A. Saifi and P. W. Smith: *Appl. Phys. Lett.* **51** (1987) 1135.

31 Y. R. Shen: Chap. 16 in *The Principles of Nonlinear Optics*, (John Wiley and Sons, Inc., New York, 1984) p. 286 and refs. cited therein.

32 B. S. Wherrett: *Nonlinear Optics: Materials and Devices*, ed. C. Flytzanis and J. L. Oudar (Springer-Verlag, Berlin, 1986) p. 180.

33 N. Peyghambarian and H. M. Gibbs: *J. Opt. Soc. Am.* **B2** (1985) 1215.

34 A. J. Glass: *Laser Program Annual Report-1974* (Lawrence Livermore Laboratory, 1975) p. 256.

35 N. L. Boling, A. J. Glass and A. Owyoung: *IEEE J. Quantum Electron*. **QE-14** (1978) 601.

36 G. Burns: Vol. 4, Chap. 3 in *Solid State Physics*, (Academic Press, Inc., 1985) (in Japanese), translated by S. Kojima, A. Sawada and T. Nakamura (Tokai Daigaku shuppankai, 1991) p. 89.

37 R. C. Miller: *Appl. Phys. Lett.* **5** (1964) 17.

38 C. C. Wang: *Phys. Rev.* **B2** (1970) 2045.

39 C. C. Wang and E. L. Baardsen: *Phys. Rev.* **185** (1969) 1079.

40 D. Milam, M. J. Weber and A. J. Glass: *Appl. Phys. Lett.* **31** (1977) 822.

41 M. J. Weber, C. F. Cline, W. L. Smith, D. Milam, D. Heiman and R. W. Hellwarth: *Appl. Phys. Lett.* **32** (1978) 403.

42 M. E. Lines: *Phys. Rev.* **B41** (1990-II) 3383.

43 M. E. Lines: *Phys. Rev.* **B43** (1991-I) 11978.

44 S. H. Kim, T. Yoko and S. Sakka: *J. Am. Ceram. Soc.* **76** (1993) 2486.

45 K. Terashima, T. Hashimoto, T. Uchino, S. Kim and T. Yokoi: *J. Ceram. Soc. Jpn.* **104** (1996) 1008.

46 K. Terashima, T. Uchino, T. Hashimoto and T. Yoko: *J. Ceram. Soc. Jpn.* **105** (1997) 288.

47 S. H. Wemple and M. DiDomenico, Jr.: *Phys. Rev.* **B3** (1971) 1338.

48 S. H. Wemple: *J. Chem. Phys.* **67** (1977) 2151.

49 R. Hellwarth, J. Cherlow and T. T. Yang: *Phys. Rev.* **B11** (1975) 964.

50 I. Kang, S. Smolorz, T. Krauss, F. Wise, B. G. Aitken and N. F. Borrelli: *Phys. Rev.* **B54** (1996-II) R12641.

51 D. Milam and M. J. Weber: *J. Appl. Phys.* **47** (1976) 2497.

52 G. Leppelmeier, W. Simmons, R. Boyd, C. Fetterman and J. E. Lane: *Laser Program Annual Report-1974* (Lawrence Livermore Laboratory, 1975) p. 160.

53 I. Thomazeau, J. Etchepare, G. Grillon and A. Migus: *Opt. Lett.* **10** (1985) 223.

54 R. Adair, L. L. Chase and S. A. Payne: *J. Opt. Soc. Am.* **B4** (1987) 875.

55 S. R. Friberg and P. W. Smith: *IEEE Quantum Electron.* **QE- 23** (1987) 2089.

56 M. A. Newhouse: *Mat. Res. Soc. Symp. Proc.* **244** (1992) 229.

57 D. W. Hall, M. A. Newhouse, N. F. Borrelli, W. H. Dumbaugh and D. L. Weidman: *Appl. Phys. Lett.* **54** (1989) 1293.

58 N. F. Borrelli, B. G. Aitken, M. A. Newhouse and D. W. Hall: *J. Appl. Phys.* **70** (1991) 2774.

59 H. Nasu, T. Uchigaki, K. Kamiya, H. Kambara and K. Kubodera: *Jpn. J. Appl. Phys.* **31** (1992) 3899.

60 F. Miyaji, K. Tadanaga and S. Sakka: *Appl. Phys. Lett.* **60** (1992) 2060.

61 E. Fargin, A. Berthereau, T. Cardinal, G. Le Flem, L. Ducasse, L. Canioni, *et al.*: *J. Non-Cryst. Solids* **203** (1996) 96.

62 L. Sarger, P. Segonds, L. Canioni, F. Adamietz, A. Ducasse, C. Duchesne, *et al.*:
    J. Opt. Soc. Am. **B11** (1994) 995.
63 R. Adair, L. L. Chase and S. A. Payne: *Phys. Rev.* **B39** (1989) 3337.
64 R. Adair, L. L. Chase and S. A. Payne: *Laser Program Annual Report-1987*
    (Lawrence Livermore Laboratory, 1989) p. 5-61.
65 I. Kang, T. D. Krauss, F. W. Wise, B. G. Aitken and N. F. Borrelli: *J. Opt. Soc. Am.*
    B12 (1995) 2053.
66 E. M. Vogel, S. G. Kosinski, D. M. Krol, J. L. Jackel, S. R. Friberg, M. K. Oliver
    et al.: *J. Non-Cryst. Solids* **107** (1989) 244.
67 E. M. Vogel, E. W. Chase, J. L. Jackel and B. J. Wilkens: *Appl. Opt.* **28** (1989) 649.
68 H. Nasu, O. Matsushita, K. Kamiya, H. Kobayashi and K. Kubodera: *J. Non-Cryst.*
    Solids **124** (1990) 275.
69 S. H. Kim, T. Yoko and S. Sakka: *J. Am. Ceram. Soc.* **76** (1993) 2486.
70 N. Sugimoto, H. Kanbara, S. Fujiwara, K. Tanaka and K. Hirao: *Opt. Lett.* **21**
    (1996) 1637.
71 R. DeSalvo, A. A. Said, D. J. Hagan, E. W. Van Stryland and M. Sheik-Bahae:
    IEEE J. Quantum Electron. **32** (1996) 1324.
72 K. Terashima, T. Hashimoto and T. Yoko: *Phys. Chem. Glass* **37** (1996) 129.
73 K. Terashima, S. H. Kim and T. Yoko: *J. Am. Ceram. Soc.* **78** (1995) 1601.
74 H. Nasu, J. Matsuoka, O. Sugimoto, M. Kida and K. Kamiya: *J. Ceram. Soc. Jpn.*
    101 (1993) 43.
75 X. Zhu, Q. Li, N. Ming and Z. Meng: *Appl. Phys. Lett.* **71** (1997) 867.
76 S. Le Boiteux, P. Segonds, L. Canioni, L. Sarger, T. Cardinal, C. Duchesne, *et al.*:
    J. Appl. Phys. **81** (1997) 1481.
77 K. Terashima, S. Tamura, S. Kim and T. Yoko: *J. Am. Ceram. Soc.* **80** (1997) 2903.
78 T. Suemura, M. Ohtani, R. Morita and M. Yamashita: *Jpn. J. Appl. Phys.* **36** (1997)
    L1307.
79 E. M. Vogel: *J. Am. Ceram. Soc.* **72** (1989) 719.
80 H. Nasu, Y. Ibara and K. Kubodera: *J. Non-Cryst. Solids* **110** (1989) 229.
81 H. Nasu, K. Kubodera, M. Kobayashi, M. Nakamura and K. Kamiya: *J. Am.*
    Ceram. Soc. **73** (1990) 1794.
82 T. I. Kosa, R. Rangel-Rojo, E. Hajto, P. J. S. Ewen, A. E. Owen, A. K. Kar, *et al.*: *J.*
    Non-Cryst. Solids **164–166** (1993) 1219.
83 M. Asobe, T. Kanamori and K. Kubodera: *IEEE Photonics Technol. Lett.* **4** (1992)
    362.
84 Z. H. Zhou, H. Nasu, T. Hashimoto and K. Kamiya: *J. Non-Cryst. Solids* **215**
    (1997) 61.
85 H. Kanbara, S. Fujiwara, K. Tanaka, H. Nasu and K. Hirao: *Appl. Phys. Lett.* **70**
    (1997) 925.
86 Z. H. Zhou, H. Nasu, T. Hashimoto and K. Kamiya: *J. Ceram. Soc. Jpn.* **105**
    (1997) 1079.
87 K. Hirao, H. Kanbara, S. Fujiwara, K. Tanaka and N. Sugimoto: *J. Ceram. Soc.*
    Jpn. **105** (1997) 1115.
88 K. A. Cerqua-Richardson, J. M. McKinley, B. Lawrence, S. Joshi and
    A. Villeneuve: *Opt. Mat.* **10** (1998) 155.
89 D. Marchese, M. De Sario, A. Jha, A. K. Kar and E. C. Smith: *J. Opt. Soc. Am.*
    B15 (1998) 2361.
90 F. Smektala, C. Quemard, L. Leneindre, J. Lucas, A. Barthelemy and C. DeAnge-
    lis: *J. Non-Cryst. Solids* **239** (1998) 139.
91 A. Owyoung, R. W. Hellwarth and N. Geo: *Phys. Rev.* **B5** (1972) 628.

92 M. Asobe and T. Kanamori: *Proc. International Workshop on Advanced Materials for Multifunctional Waveguides*, Chiba (1995) p. 64.

93 M. Asobe, T. Kanamori and K. Kubodera: *IEEE J. Quantum Electron.* **29** (1993) 2325.

94 M. Asobe, H. Itoh, T. Miyazawa and T. Kanamori: *Electron. Lett.* **29** (1993) 1966.

95 E. Hanamura: *Optical and Quantum Electron.* **21** (1989) 441 and refs. cited therein.

96 L. Brus: *IEEE J. Quantum Electron.* **QE-22** (1986) 1909.

97 D. S. Chemla, D. A. B. Miller and S. Schmitt-Rink: Chap. 4 in *Optical Nonlinearities and Instabilities in Semiconductors*, ed. H. Haug, (Academic Press, Inc., Boston, 1988) p. 83.

98 S. Schmitt-Rink, S. A. B. Miller and D. S. Chemla: *Phys. Rev.* **B35** (1987-II) 8113.

99 H. A. Pohl: *Quantum Mechanics For Science and Engineering*, (Prentice-Hall, Inc., Englewood Cliffs, 1967), Jpn. trans. A. Tsugawa (Uchida Roukakuho, Tokyo, 1976).

100 J. C. Phillips: *Bonds and Bands in Semiconductors*, (Academic Press, Inc., New York, 1973) p. 8.

101 C. Kittel: *Introduction to Solid State Physics*, 6th edn, Chap. 6 and 7 (John Wiley & Sons, Inc., New York, 1986) p. 127.

102 A. J. Dekker: Chap. 10 in *Solid State Physics*, (Prentice-Hall, Inc., Englewood Cliffs, Maruzen Company, Ltd., Tokyo, 1957) p. 238.

103 R. A. Smith: Chap. 2 in *Semiconductors* 2nd edn, (Cambridge University Press, Cambridge, 1986) p. 24.

104 C. Kittel: Chap. 11 in *Introduction to Solid State Physics*, 6th edn, (John Wiley & Sons, Inc., New York, 1986) p. 253.

105 R. A. Smith: Chap. 3 in *Semiconductors*, 2nd edn, (Cambridge University Press, Cambridge, 1986) p. 46.

106 R. S. Knox: *Theory of Excitons*, Solid State Physics, Supplement Vol. 5, Chap. II (Academic Press, New York, 1963) p. 7.

107 Al. L. Efros and A. L. Efros: *Sov. Phys. Semicond.* **16** (1982) 772.

108 A. I. Ekimov and A. A. Onushchenko: *Sov. Phys. Semicond.* **16** (1982) 775.

109 A. I. Ekimov and A. A. Onushchenko: *JETP Lett.* **40** (1984) 1136.

110 A. I. Ekimov, Al. L. Efros and A. A. Onushchenko: *Solid State Commun.* **56** (1985) 921.

111 L. E. Brus: *J. Chem. Phys.* **80** (1984) 4403.

112 H. Shinojima, J. Yumoto, N. Uesugi, S. Ohmi and Y. Asahara: *Appl. Phys. Lett.* **55** (1989) 1519.

113 N. F. Borrelli, D. W. Hall, H. J. Holland and D. W. Smith: *J. Appl. Phys.* **61** (1987) 5399.

114 B. G. Potter, Jr and J. H. Simmons: *Phys. Rev.* **B37** (1988- II) 10838.

115 T. Tokizaki, T. Kataoka, A. Nakamura, N. Sugimoto and T. Manabe: *Jpn. J. Appl. Phys.* **32** (1993) L 782.

116 S. C. Hsu and H. S. Kwok: *Appl. Phys. Lett.* **50** (1987) 1782.

117 L. H. Acioli, A. S. L. Gomes, J. R. R. Leite and Cid. B. de Araujo: *J. Quantum Electron.* **26** (1990) 1277.

118 P. Roussignol, M. Kull, D. Ricard, F. de Rougemont, R. Frey and C. Flytzanis: *Appl Phys. Lett.* **51** (1987) 1882.

119 C. N. Ironside, T. J. Cullen, B. S. Bhumbra, J. Bell, W. C. Banyai, N. Finlayson, et al.: *J. Opt. Soc. Am.* **B5** (1988) 492.

120 A. Gabel, K. W. Delong, C. T. Seaton and G. I. Stegeman: *Appl. Phys. Lett.* **51** (1987) 1682.

121 S. S. Yao, C. Karaguleff, A. Gebel, R. Fortenberry, C. T. Seaton and G. I. Stegeman: *Appl. Phys. Lett.* **46** (1985) 801.

122 P. Roussignol, D. Richard and C. Flytzanis: *Appl. Phys.* **A44** (1987) 285.

123 D. Cotter: *Electron. Lett.* **22** (1986) 693.

124 J. T. Remillard and D. G. Steel: *Opt. Lett.* **13** (1988) 30.

125 B. Vaynberg, M. Matusovsky, M. Rosenbluh, E. Kolobkova and A. Lipovskii: *Opt. Commun.* **132** (1996) 307.

126 J. T. Remillard, H. Wang, M. D. Webb and D. G. Steel: *IEEE Quantum Electron.* **25** (1989) 408.

127 M. C. Nuss, W. Zinth and W. Kaiser: *Appl. Phys. Lett.* **49** (1986) 1717.

128 S. M. Saltiel, B. Van Wonterghem and P. M. Rentzepis: *Opt. Lett.* **14** (1989) 183.

129 J. Yumoto, S. Fukushima and K. Kubodera: *Opt. Lett.* **12** (1987) 832.

130 B. Vaynberg, M. Matusovsky, M. Rosenbluh, V. Petrikov and A. Lipovskii: *J. Appl. Phys.* **81** (1997) 6934.

131 L. H. Acioli, A. S. L. Gomes and J. R. Rios Leite: *Appl. Phys. Lett.* **53** (1988) 1788.

132 F. de Rougemont, R. Frey, P. Roussignol, D. Ricard and C. Flytzanis: *Appl. Phys. Lett.* **50** (1987) 1619.

133 P. Roussignol, D. Ricard, J. Lukasik and C. Flytzanis: *J. Opt. Soc. Am.* **B4** (1987) 5.

134 D. Ricard, P. Roussignol, F. Hashe and Ch. Flytzanis: *phys. stat. sol.* (b) **159** (1990) 275.

135 T. Hiroshima and E. Hanamura: *Solid State Physics*, Special Issue, Optical Nonlinear Materials **24** (1989) 839 (in Japanese).

136 Ph. Roussignol, D. Ricard, Chr. Flytzanis: *Appl. Phys.* **B 51** (1990) 437.

137 Y. Asahara: *Advances in Science and Technology*, Vol. 4, *New Horizons for Materials*, ed. P. Vincenzini (Techna Srl, 1995) p. 91.

138 Y. Asahara: *Ceramic International*, **23** (1997) 375.

139 P. Roussignol, D. Ricard, K. C. Rustagi and C. Flytzanis: *Opt. Commun.* **55** (1985) 143.

140 S. H. Park, R. A. Morgan, Y. Z. Hu, M. Lindberg, S. W. Koch and N. Peyghanbarian: *J. Opt. Soc. Am.* **B7** (1990) 2097.

141 D. W. Hall and N. F. Borrelli: *J. Opt. Soc. Am.* **5** (1988) 1650.

142 R. A. Smith: Chap. 3 in *Semiconductors*, 2nd edn, (Cambridge University Press, Cambridge, 1986) p. 264.

143 N. Payghambarian and H. M. Gibbs: Semiconductor Optical Nonlinearities and Applications of Optical Devices and Bistability, in *Optical Nonlinearities and Instabilities in Semiconductors*, ed. H. Haug, (Academic Press, Inc., Boston, 1988) p. 295.

144 J. Warnock and D. D. Awschalom: *Appl. Phys. Lett.* **48** (1986) 425.

145 J. Warnock and D. D. Awschalom: *Phys. Rev.* **B32** (1985) 5529.

146 C. Hata, H. Hiraga, S. Ohmi, K. Uchida, Y. Asahara, A. J. Ikushima, *et al.*: *Technical Digest of Optical Society of America, 1991* Annual Meeting, San Jose (Optical Society of America, Washington, DC, 1991) MSS-4.

147 M. Kull, J. L. Coutaz, G. Manneberg and V. Grivickas: *Appl. Phys. Lett.* **54** (1989) 1830.

148 W. C. Banyai, N. Finlayson, C. T. Seaton, G. I. Stegeman, M. O'Neill, T. J. Cullen *et al.*: *Appl. Phys. Lett.* **54** (1989) 481.

149 N. Peyghambarian, B. Fluegel, D. Hulin, A. Migus, M. Joffre, A. Antonetti, *et al.*: *IEEE J. Quantum Electron.* **25** (1989) 2516.

150 I. V. Tomov and P. M. Rentzepis: *J. Opt. Soc. Am.* **B9** (1992) 232.

151 J. P. Zheng, L. Shi, F. S. Choa, P. L. Liu and H. S. Kwok: *Appl. Phys. Lett.* **53** (1988) 643.

152 M. Tomita, T. Matsumoto and M. Matsuoka: *J. Opt. Soc. Am.* **B6** (1989) 165.

153 F. Hache, M. C. Klein, D. Ricard and C. Flytzanis: *J. Opt. Soc. Am.* **B8** (1991) 1802.

154 M. C. Schanne-Klein, F. Hache, D. Ricard and C. Flytzanis: *J. Opt. Soc. Am.* **B9** (1992) 2234.

155 J. T. Remillard, H. Want, M. D. Webb and D. G. Steel: *J. Opt. Soc. Am.* **B7** (1990) 897.

156 J. P. Zheng and H. S. Kwok: *J. Opt. Soc. Am.* **B9** (1992) 2047.

157 K. Natterman, B. Danielzik and D. von der Linde: *Appl. Phys.* **A44** (1987) 111.

158 M. Ghanassi, M. C. Schanne-Klein, F. Hache, A. I. Ekimov, D. Ricard and C. Flytzanis: *Appl. Phys. Lett.* **62** (1993) 78.

159 M. Kull and J. L. Coutaz: *J. Opt. Soc. Am.* **B7** (1990) 1463.

160 K. W. Delong, A. Gabel, C. T. Seaton and G. I. Stegeman: *J. Opt. Soc. Am.* **B6** (1989) 1306.

161 V. I. Klimov and D. W. McBranch: *Phys. Rev.* **B55** (1997-I) 13173.

162 V. I. Klimov, P. H. Bolivar and H. Kurz: *Phys. Rev.* **B53** (1996-I) 1463.

163 M. Mitsunaga, H. Shinogima and K. Kubodera: *J. Opt. Soc. Am.* **B5** (1988) 1448.

164 B. Van Wonterghem, S. M. Saltiel, T. E. Dutton and P. M. Rentzepis: *J. Appl. Phys.* **66** (1989) 4935.

165 P. Horan and W. Blau: *J. Opt. Soc. Am.* **B7** (1990) 304.

166 A. Vanhaudenarde, M. Trespidi and R. Frey: *J. Opt. Soc. Am.* **B11** (1994) 1474.

167 H. Ma, A. S. L. Gomes and Cid B. de Araujo: *J. Opt. Soc. Am.* **B9** (1992) 2230.

168 J. Malhotra, D. J. Hagan and B. G. Potter: *J. Opt. Soc. Am.* **B8** (1991) 1531.

169 M. Tomita and M. Matsuoka: *J. Opt. Soc. Am.* **B7** (1990) 1198.

170 H. Hiraga, C. Hata, K. Uchida, S. Ohmi, Y. Asahara, A. J. Ikushima, *et al.*: *XVI Int. Congress on Glass*, ed. D. E. de Ceramica y Vidrio, **Vol. 3** (1992) p. 421.

171 P. Maly, F. Trojanek and A. Svoboda: *J. Opt. Soc. Am.* **B10** (1993) 1890.

172 M. Wittmann and A. Penzkofer: *Opt. Quantum Electron.* **27** (1995) 705.

173 H. Shinojima, J. Yumoto, N. Uesugi, S. Ohmi and Y. Asahara: Extended abstracts of *Topical Meeting on Glasses for Optoelectronics* (The Ceramic Society of Japan, International Commission on Glass, Tokyo, 1989) p. 73.

174 H. Okamoto, J. Matsuoka, H. Nasu, K. Kamiya and H. Tanaka: *J. App. Phys.* **75** (1994) 2251.

175 Y. Liu, V. C. S. Reynoso, L. C. Barbosa, R. F. C. Rojas, H. L. Fragnito, C. L. Cesar *et al.*: *J. Mat. Sci. Lett.* **14** (1995) 635.

176 S. Omi, H. Hiraga, K. Uchida, C. Hata, Y. Asahara, A. J. Ikushima, *et al.*: *Proc. Int. Conf. on Science and Technology of New Glasses*, ed. S. Sakka and N. Soga (1991) p. 181.

177 U. Woggon, M. Müller, I. Rückmann, J. Kolenda and M. Petrauskas: *phys. stat. sol. (b)* **160** (1990) K79.

178 T. Yanagawa, H. Nakano, Y. Ishida and Y. Sasaki: *Opt. Commun.* **88** (1992) 371.

179 E. Hanamura: *The Physics of Non-Crystalline Solids*, ed. L. D. Pye, W. C. La Course and H. J. Stevens, (The Society of Glass Technology, Taylor and Francis, London, 1992) p. 528.

180 E. Hanamura: *phys. stat. sol. (b)* **173** (1992) 241.

181 I. Broser and J. Gutowski: *Appl. Phys.* **B46** (1988) 1.
182 B. L. Justus, M. E. Seaver, J. A. Ruller and A. J. Campillo: *Appl. Phys. Lett.* **57** (1990) 1381.
183 L. G. Zimin, S. V. Gaponenko, V. Yu. Lebed, I. E. Malinovskii, I. N. Germanenko, E. E. Podorova *et al.*: *phys. stat. sol. (b)* **159** (1990) 267.
184 P. Gilliot, B. Hönerlage, R. Levy and J. B. Grun: *phys. stat. sol. (b)* **159** (1990) 259.
185 P. Gilliot, J. C. Merle, R. Levy, M. Robino and B. Hönerlage: *phys. stat. sol. (b)* **153** (1989) 403.
186 U. Woggon, F. Henneberger and M. Müller: *phys. stat. sol. (b)* **150** (1988) 641.
187 F. Henneberger, U. Woggon, J. Puls and Ch. Spiegelberg: *Appl. Phys.* **B46** (1988) 19.
188 A. Nakamura, H. Yamada and T. Tokizaki: *Phys. Rev.* **B40** (1989-II) 8585.
189 T. Kataoka, T. Tokizaki and A. Nakamura: *Phys. Rev.* **B48** (1993-II) 2815.
190 Y. Li, M. Takata and A. Nakamura: *Phys. Rev.* **B57** (1998-I) 9193.
191 Y. Li, M. Ohta, S. Sasaki and A. Nakamura: *Jpn. J. Appl. Phys.* **37** (1998) L33.
192 Y. Li, M. Ohta and A. Nakamura: *Phys. Rev.* **B57** (1998-II) R12673.
193 W. Vogel: *Chemistry of Glass*, translated and edited by N. Kreidle, Chap. 8, (The American Ceramic Society, Inc., Columbus, 1985) p. 153.
194 J. Yumoto, H. Shinojuma, N. Uesugi, K. Tsunetomo, H. Nasu and Y. Osaka: *Appl. Phys. Lett.* **57** (1990) 2393.
195 T. Takada, T. Yano, A. Yasumori, M. Yamane and J. D. Mackenzie: *J. Non-Cryst. Solids* **147 & 148** (1992) 631.
196 M. Nogami, Y.-Q. Zhu, Y. Tohyama, K. Nagasaka, T. Tokizaki and A. Nakamura: *J. Am. Ceram. Soc.* **74** (1991) 238.
197 M. D. Dvorak, B. L. Justus, D. K. Gaskill and D. G. Hendershot: *Appl. Phys. Lett.* **66** (1995) 804.
198 D. C. Rogers, R. J. Manning, B. J. Ainslie, D. Cotter, M. J. Yates, J. M. Parker, *et al.*: *IEEE Photonics Technol. Lett.* **6** (1994) 1017.
199 W.-T. Han and Y. K. Yoon: *J. Non-Cryst. Solids* **196** (1996) 84.
200 S. Omi, H. Hiraga, K. Uchida, C. Hata, Y. Asahara, A. J. Ikushima, T. Tokizaki and A. Nakamura: *Tech. Digest Series* Vol. 10, *Conf. on Lasers and Electro-optics*, Baltimore (Optical Society of America, 1991) p. 88.
201 A. Tabata, N. Matsuno, Y. Suzuoki and T. Mizutani: *Jpn. J. Appl. Phys.* **35** (1996) 2646.
202 N. Sugimoto, M. Yamamoto, T. Manabe, S. Ito, T. Tokizaki, T. Kataoka and A. Nakamura: *Proc. Int. Conf. on Science and Technology of New Glasses* (Ceramic Society of Japan, Tokyo, 1991) p. 394.
203 S. Ohtsuka, K. Tsunetomo, T. Koyama and S. Tanaka: *Extended Abstract, International Symposium on Non-linear Photonics Materials*, ed. Japan Industrial Technology Association (Japan High Polymer Center, 1994) p. 205.
204 V. Klimov, P. H. Bolivar, H. Kurz, V. Karavanskii, V. Krasovskii and Yu. Korkishko: *Appl. Phys. Lett.* **67** (1995) 653.
205 V. Klimov and V. A. Karavanskii: *Phys. Rev.* **B54** (1996-I) 8087.
206 B. Yu, C. Zhu, H. Xia, H. Chen and F. Gan: *Mat. Sci. Lett.* **16** (1997) 2001.
207 Y. Kuroiwa, N. Sugimoto, S. Ito and A. Nakamura: Optical Society of America, '98 Annual Meeting, Baltimore (1998) MV5.
208 N. Finlayson, W. C. Banyai, E. M. Wright, C. T. Seaton, G. I. Stegeman, T. J. Cullen *et al.*: *Appl. Phys. Lett.* **53** (1988) 1144.

209 P. T. Guerreiro, S. Ten, N. F. Borrelli, J. Butty, G. E. Jabbour and N. Peyghambarian: *Appl. Phys. Lett.* **71** (1997) 1595.

210 F. Abeles: Chap. 3 in *Optical Properties of Solids*, ed. F. Abeles (North-Holland Publishing Company, Amsterdam, London, 1972) p. 93.

211 C. Kittel: Chap. 12 in *Introduction to Solid State Physics*, (2nd edn) (John Wiley and Sons, Inc., New York, Maruzen Company Ltd., Tokyo, 1956) p. 312.

212 M. P. Givens: Optical Properties of Metals in *Solid State Physics*, Vol. 6, ed. F. Seitz and D. Turnbull, (Academic Press, New York, 1958) p. 313.

213 J. A. A. J. Perenboom, P. Wyder and F. Meier: *Electronic Properties of Small Metallic Particles*, Physics Report (Review Section of *Physics Letters*) **78** (1981) 173.

214 U. Kreibig and C. V. Fragstein: *Z. Physik* **224** (1969) 307.

215 C. F. Bohren and D. R. Huffman: *Absorption and Scattering of Light by Small Particles*, (John Wiley & Sons, Inc., 1983).

216 W. P. Halperin: Quantum size effects in metal particles. *Rev. Modern Phys.* **58** (1986) 533.

217 G. C. Papavassiliou: Optical properties of small inorganic and organic metal particles, *Prog. Solid St. Chem.* **12** (1979) 185.

218 D. Ricard: Nonlinear Optics at Surface and in Composite Materials, in *Nonlinear Optics: Materials and Devices*, ed. C. Flytzanis and J. L. Oudar (Springer-Verlag, Berlin, 1985) p. 154.

219 J. C. Maxwell-Garnett: *Phil. Trans. R. Soc. London* **203** (1904) 385.

220 J. C. Maxwell-Garnett: *Phil. Trans. R. Soc. London* **205** (1906) 237.

221 R. H. Doremus: *J. Chem. Phys.* **40** (1964) 2389.

222 R. Doremus, S. C. Kao and R. Garcia: *Appl. Opt.* **31** (1992) 5773.

223 D. Ricard, Rh. Roussignol and Chr. Flytzanis: *Opt. Lett.* **10** (1985) 511.

224 F. Hache, D. Ricard, C. Flytzanis and U. Kreibig: *Appl. Phys.* **A47** (1988) 347.

225 F. Hache, D. Ricard and C. Flytzanis: *J. Opt. Soc. Am.* **B3** (1986) 1647.

226 Li Yang, K. Becker, F. M. Smith, R. H. Magruder III, R. F. Haglund, Jr., Lina Yang, *et al.*: *J. Opt. Soc. Am.* **B11** (1994) 457.

227 L. Yang, D. H. Osborne, R. F. Haglund, Jr., R. H. Magruder, C. W. White, R. A. Zuhr *et al.*: *Appl. Phys.* **A62** (1996) 403.

228 K. Uchida, S. Kaneko, S. Omi, C. Hata, H. Tanji, Y. Asahara, *et al.*: *J. Opt. Soc. Am.* **B11** (1994) 1236.

229 J. Matsuoka, R. Mizutani, S. Kaneko, H. Nasu, K. Kamiya, K. Kadono, *et al.*: *J. Ceram. Soc. Jpn.* **101** (1993) 53.

230 M. Mennig, M. Schmitt, U. Becker, G. Jung and H. Schmidt: *SPIE* Vol. **2288** Sol-Gel Optics III, (1994) p. 130.

231 M. Lee, T. S. Kim and Y. S. Choi: *J. Non-Cryst. Solids* **211** (1997) 143.

232 Y. Hosoya, T. Suga, T. Yanagawa and Y. Kurokawa: *J. Appl. Phys.* **81** (1997) 1475.

233 M. J. Bloemer, J. W. Haus and P. R. Ashley: *J. Opt. Soc. Am.* **B7** (1990) 790.

234 R. W. Schoenlein, W. Z. Lin, J. G. Fujimoto and G. L. Eesley: *Phys. Rev. Lett.* **58** (1987) 1680.

235 R. H. Magruder III, R. F. Haglund, Jr., L. Yang, J. E. Wittig, K. Becker and R. A. Zuhr: *Mat. Res. Soc. Symp. Proc.* **224** (Materials Research Society, 1992) 369.

236 T. Tokizaki, A. Nakamura, S. Kaneko, K. Uchida, S. Omi, H. Tanji *et al.*: *Appl. Phys. Lett.* **65** (1994) 941.

237 M. Perner, P. Bost, U. Lemmer, G. von Plessen, J. Feldmann, U. Becker, *et al.*: *Phys. Rev. Lett.* **78** (1997) 2192.

238 Y. Hamanaka and A. Nakamura: *Abstracts of the Meeting of the Physical Society of Japan*, Sectional Meeting Part 2 (1996) p. 358, 4p-E-4.

239 Y. Hamanaka, N. Hayashi and A. Nakamura: *Abstract of the 52 Annual Meeting of the Physical Society of Japan*, Part 2 (1997) p. 227, 30p-YC-9.

240 Y. Hamanaka, N. Hayashi and A. Nakamura: *Meeting Abstract of Sectional Meeting of the Physical Society of Japan*, Part 2 (1997) p. 243, 7p-J-7.

241 T. Kineri, M. Mori, K. Kadono, T. Sakaguchi, M. Miya, H. Wakabayashi *et al.*: *J. Ceram. Soc. Jpn.* **101** (1993) 1340.

242 T. Kineri, M. Mori, K. Kadono, T. Sakaguchi, M. Miya, H. Wakabayashi *et al.*: *J. Ceram. Soc. Jpn.* **103** (1995) 117.

243 I. Tanahashi, Y. Manabe, T. Tohda, S. Sasaki and A. Nakamura: *J. Appl. Phys.* **79** (1996) 1244.

244 K. Uchida, S. Kaneko, S. Omi, C. Hata, H. Tanji, Y. Asahara, *et al.*: *Optical Society of America, Annual Meeting*, Dallas (1994) Tu 14.

245 P. B. Johnson and R. W. Christy: *Phys. Rev.* **B6** (1972) 4370.

246 H. Wakabayashi, H. Yamanaka, K. Kadono, T. Sakaguchi and M. Miya: *Proc. Int. Conf. on Science and Technology of New Glass*, (Ceramic Society of Japan, Tokyo, 1991) p. 412.

247 K. Fukumi, A. Chayahara, K. Kadono, T. Sakaguchi, Y. Horino, M. Miya, *et al.*: *Jpn. J. Appl. Phys.* **30** (1991) L742.

248 R. H. Magruder III, Li Yang, R. F. Haglund, Jr., C. W. White, Lina Yang, R. Dorsinville, *et al.*: *Appl. Phys. Lett.* **62** (1993) 1730.

249 K. Kadono, T. Sakaguchi, H. Wakabayashi, T. Fukimi, H. Yamanaka, M. Miya and H. Tanaka: *Mat. Res. Soc. Symp. Proc. 282* (Materials Research Society, 1993) p. 903.

250 R. F. Haglund, Jr., L. Yang, R. H. Magruder, III, J. E. Wittig, K. Becker and R. A. Zuhr: *Opt. Lett.* **18** (1993) 373.

251 T. Akai, K. Kadono, H. Yamanaka, T. Sakaguchi, M. Miya and H. Wakabayashi: *J. Ceram. Soc. Jpn.* **101** (1993) 105.

252 K. Fukumi, A. Chayahara, K. Kadono, T. Sakaguchi, Y. Horino, M. Miya, *et al.*: *J. Appl. Phys.* **75** (1994) 3075.

253 H. Hosono, Y. Abe, Y. L. Lee, T. Tokizaki and A. Nakamura: *Appl. Phys. Lett.* **61** (1992) 2747.

254 R. H. Magruder, III, D. H. Osborne, Jr. and R. A. Zuhr: *J.. Non-Cryst. Solids* **176** (1994) 299.

255 Y. Takeda, T. Hioki, T. Motohiro and S. Noda: *Appl. Phys. Lett.* **63** (1993) 3420.

256 I. Tanahashi, M. Yoshida, Y. Manabe, T. Tohda, S. Sasaki, T. Tokizaki *et al.*: *Jpn. J. Appl. Phys.* **33** (1994) L1410.

257 R. H. Magruder, III, R. F. Haglund, Jr., L. Yang, J. E. Wittig and R. A. Zuhr: *J. Appl. Phys.* **76** (1994) 708.

258 G. De Marchi, F. Gonella, P. Mazzoldi, G. Battaglin, E. J. Knystautas and C. Meneghini: *J. Non-Cryst. Solids* **196** (1996) 79.

259 H. B. Liao, R. F. Xiao, J. S. Fu, P. Yu, G. K. L. Wong and Ping Sheng: *Appl. Phys. Lett.* **70** (1997) 1.

260 H. B. Liao, R. F. Xiao, J. S. Fu, H. Wang, K. S. Wong and G. K. L. Wong: *Opt. Lett.* **23** (1998) 388.

261 Ji-Hye Gwak, L. Chae, S.-J. Kim and M. Lee: *Nano Structured Materials*, **8** (1997) 1149.

262 M. Falconieri, G. Salvetti, E. Cattaruzza, R. Gonella, G. Mattei, P. Mazzoldi, *et al.*: *Appl. Phys. Lett.* **73** (1998) 288.

263 E. Cattaruzza, F. Gonella, G. Mattei, P. Mazzoldi, D. Gatteschi, C. Sangregorio, *et al.*: *Appl. Phys. Lett.* **73** (1998) 1176.

264 N. Del Fatti, F. Vallee, C. Flytzanis, Y. Hamanaka, A. Nakamura and S. Omi: *Technical Digest. Int. Quantum Electronics Conf. (IQEC)*, San Francisco, May 3–8, (Optical Society of America, Washington, DC, 1998) p. 93. QWC3.

265 H. Nakanishi: Extended Abstracts, The Sixth Symposium on Nonlinear Photonics Materials (Japan Chemical Innovation Institute, 1999) p. 1.

5

Magneto-optical glass

Introduction

The research into materials with a high magneto-optical effect is always an interesting matter because these materials are extensively used as magnetic field sensors and optical isolators. Glass materials are of interest for these applications because they are transparent in the visible and near infrared spectral region and can be readily formed into complex shapes such as optical fibers. One of the magneto-optical effects, the Faraday effect in glass is a well-known phenomenon. The primary trend of the studies is, therefore, to develop and produce glass compositions having a large specific Faraday rotation and low absorption in the visible and near infrared regions.

Generally the Verdet constant, a measure of the Faraday rotation of the materials, is considered to be of two types depending upon the ion or ions that are incorporated in glass: diamagnetic or paramagnetic. Most normal network former and modifier ions in glass would give rise to diamagnetic rotation. The rare-earth and transition ions are examples of paramagnetic ions. Diamagnetic glasses generally have small and positive Verdet constants, which are almost independent of temperature, whereas paramagnetic glasses usually have large and negative Verdet constants, which are generally inversely proportional to temperature. The Faraday effects of various glasses are presented and discussed in this chapter.

5.1 Magnetic properties of materials

5.1.1 Origin of magnetics of materials

5.1.1.1 Permanent magnetic dipoles [1, 2]

From the hypothesis of Ampere, the magnetic dipole moment associated with the electron orbit radius r and the angular velocity ω_0 is given by

$$\mu_m = -e\omega_0 r^2/2c. \tag{5.1}$$

Since the angular momentum of the electron is given by $\gamma = m\omega_0 r^2$, the magnetic dipole moment can be related to the angular momentum as

$$\mu_m = -(e/2mc)\gamma. \tag{5.2}$$

According to the quantum theory, on the other hand, the angular momentum of the electron orbit is determined by the quantum number l, which is restricted to the set of values $l = 0, 1, 2, \ldots, (n-1)$. By using the angular momentum $\hbar l$, the magnetic dipole moment is given by

$$\boldsymbol{\mu}_l = -(e\hbar/2mc)\boldsymbol{l} = -\mu_B \boldsymbol{l}. \tag{5.3}$$

Here, the quantity $\mu_B = (e\hbar/2mc) = 0.927 \times 10^{-20}$ erg/Oe (emu) is called the Bohr magneton, the number of which is a measure of the magnetic moment. The electron itself has another angular momentum s known as the spin. The magnetic moment component $\boldsymbol{\mu}_s$ of the spin is given by

$$\boldsymbol{\mu}_s = -(e/2mc)(\hbar/2)\boldsymbol{s} = -2\mu_B \boldsymbol{s}. \tag{5.4}$$

In atoms containing a number of electrons, the orbital angular momentum may be combined to form a resultant $\boldsymbol{L}$ and the spin may be combined vectorially to form a resultant $\boldsymbol{S}$. Then the magnetic dipole moment associated with the total orbital angular momentum $\boldsymbol{L}$ and total spin angular momentum $\boldsymbol{S}$ is respectively

$$\boldsymbol{\mu}_L = -\mu_B \boldsymbol{L} \tag{5.5}$$
$$\boldsymbol{\mu}_S = -2\mu_B \boldsymbol{S}. \tag{5.6}$$

The total magnetic dipole moment of the whole electron system of the atom in free space is given by

$$\begin{aligned}\boldsymbol{\mu}_J &= (\boldsymbol{\mu}_L + \boldsymbol{\mu}_S) \\ &= -\mu_B(\boldsymbol{L} + 2\boldsymbol{S}) \\ &= -\mu_B g \boldsymbol{J}. \end{aligned} \tag{5.7}$$

A quantity g, called the Lande g factor or the spectroscopic splitting factor, is given by the Lande equation for a free atom

$$g = 1 + \{J(J+1) + S(S+1) - L(L+1)\}/2J(J+1). \tag{5.8}$$

For an electron spin, g is usually taken as 2.00.

5.1.1.2 Effect of a magnetic field on an atom [1]

According to the quantum theory, when the atom is subjected to an external magnetic field, the permanent magnetic moment of a given atom or ion is rotating and the quantum states of an atom are in general modified, i.e. states which were formerly degenerate may become separated and certain spectral lines may consequently be split into several components. This effect is called Zeeman splitting. The modification is, however, not free but restricted to a finite set of orientations relative to the applied field. The possible components of the magnetic moment along the direction of an external magnetic field H are determined by the magnetic quantum number m, where m is restricted to the set of values

$$m = l, (l-1), \ldots, 0, \ldots, -(l-1), -l.$$

The total orbit momentum quantum number L or the total angular momentum quantum number J of each atom in a magnetic field, also gives rise to the possible components of the magnetic moment, $M_L \mu_B$ or $M_J g \mu_B$, respectively, where

$$M_L = L, (L-1), \ldots, -(L-1), -L,$$
$$M_J = J, (J-1), \ldots, -(J-1), -J.$$

The energy of the system in a magnetic field, for instance, associated with J is, therefore, given by

$$E = M_J g \mu_B H \tag{5.9}$$

with equal energy space ΔE, which is given by

$$\Delta E = g \mu_B H. \tag{5.10}$$

5.1.2 Macroscopic description of the magnetic properties of materials [1, 2]

5.1.2.1 Magnetization

In the absence of an applied magnetic field, the average magnetic moment magnetization is zero. The application of the magnetic field H will cause the orientation of the permanent magnetic moment μ, resulting in the so-called magnetization M. The magnetism requires, therefore, the existence of permanent magnetic dipoles. The magnetization is given by

$$M = \chi H \tag{5.11}$$

where χ is the macroscopic magnetic susceptibility.

5.1.2.2 Diamagnetic materials

An atom with filled electron shells has zero spin, zero orbital moment and therefore zero permanent magnetic moment. When the flux through an electrical circuit is changed, current according to the familiar Lenz's law is induced in such a direction as to oppose the flux change. The application of the magnetic field will cause a finite current around the nucleus, resulting in a magnetic moment that is opposite to the applied magnetic field. The magnetic moment associated with such an induced current is called a diamagnetic moment.

The diamagnetic susceptibility per unit volume is, if N is the number of atoms per unit volume and Z is the number of electrons per atom,

$$\chi = -(NZe^2/6mc^2)\langle r^2 \rangle \tag{5.12}$$

where $\langle r^2 \rangle$ is the mean square distance of the electrons from the nucleus. The diamagnetic susceptibility is thus determined essentially by the charge distribution in the atoms. The absolute magnitude of diamagnetic susceptibility therefore increases with the number of electrons per ion.

Now, let us consider the influence of magnetic field on the motion of an electron in an atom quantitatively. The magnetic field H produces the Lorentz force $(e/c)vH$ on the electron, so that, according to Newtonian mechanics, we may write

$$mr\omega^2 = F - (e/c)vH$$
$$= mr\omega_0^2 - er\omega H/c \tag{5.13}$$
$$\omega^2 - \omega_0^2 = -e\omega H/mc \tag{5.14}$$

where ω is the angular velocity of the electron in a magnetic field. Assuming $\omega \sim \omega_0$, then

$$\omega^2 - \omega_0^2 = 2\omega(\omega - \omega_0) \tag{5.15}$$

and

$$\omega - \omega_0 = \omega_L$$
$$= -(e/2mc)H. \tag{5.16}$$

This means that the motion of the electron in a magnetic field is just the same as a motion in the absence of H except for the superposition of a common precession of angular frequency ω_L. The frequency ω_L is called 'Larmor frequency'. The changes produced in the energy levels by a magnetic field can be regarded as

$$\Delta E = \hbar\omega_L = -\hbar(e/2mc)H = -\mu_B H. \tag{5.17}$$

5.1.2.3 Paramagnetic materials

As was noted from Eq. (5.9), the possible component of the permanent magnetic moment is $M_J g \mu_B$. In the absence of an applied magnetic field, the average magnetic moment is zero. In an applied magnetic field, magnetization results from the orientation of the permanent magnetic moment. According to statistical mechanisms, the resultant magnetization must be given by summing the $(2J + 1)$ magnetic moment components of the corresponding M_J states per atom along H. However, thermal distortion resists the tendency of the field to orient the magnetic moment. The magnetic susceptibility in the limit, $M_J g \mu_B H / kT \ll 1$, is then given by

$$\chi \sim NJ(J + 1)g^2 \mu_B^2 / 3kT = N\mu_{\text{eff}}^2 / 3kT \tag{5.18}$$

$$\mu_{\text{eff}} = g[J(J + 1)]^{1/2} \mu_B = n_{\text{eff}} \mu_B, \tag{5.19}$$

where μ_{eff} is the statistical average of the magnetic moment and is called the effective magnetic moment, and n_{eff} is called the effective number of the Bohr magneton. The magnetism associated with this magnetization is called paramagnetism.

Paramagnetic ions in general have unfilled inner electronic shells. In the rare earth group, the unoccupied 4f electron shell lies relatively deep inside the ions because the outer electrons occupy 5s and 5p levels, thus leaving the orbits of the 4f electrons practically the same as in the free ion. The effective number of Bohr magnetons calculated from the J and g values predicts the behaviour of most of the rare-earth salts quite well except the ions Sm^{3+} and Eu^{3+}.

5.2 Fundamentals of the Faraday effect

5.2.1 Origin of Faraday rotation [3–6]

The Faraday effect is the rotation of the plane of polarization of linearly polarized light propagating through a glass parallel to an applied magnetic field. This phenomenon was discovered by M. Faraday in 1830 in PbO-containing flint glasses [3, 4] and has been utilized in the constitution of optical switches, modulators, circulators, isolators and sensors for magnetic fields and electric currents. The angle of rotation (θ) is proportional to the light path (d) through the medium and to the magnetic field strength (H), so that

$$\theta = VdH \tag{5.20}$$

where the constant of proportionality V is known as the Verdet constant (min/Oe-cm or rad/T-m) [V(min/Oe-cm) $= 3.44 \times 10^{-3} V$ (rad/T-m)], depending on the magnetic properties of the glass.

The Faraday effect originates from the difference between the refractive indices for the right- and left-circularly polarized lights, i.e. n^+ and n^- with applied longitudinal magnetic field. Plane-polarized light may be thought of as a superposition of two contra-rotating circularly polarized lights. In the presence of a magnetic field, these lights are propagated through a medium with different velocities, corresponding to the two refractive indices n^+ and n^-. Consequently, the resultant plane of recombined polarization light will be rotated at an angle θ with respect to that of the original plane. For a sample of thickness d, this angle is macroscopically expressed by [5, 6]

$$\theta = (\omega d/2c)(n^- - n^+) \tag{5.21}$$

where θ is the angle of rotation per unit length. Here θ is defined to be positive when it is clockwise looking in the direction of propagation for a magnetic field in the same direction.

In terms of the classical description, the equations of motion of electrons in the equilibrium position by the applied optical field and by the applied magnetic field includes the Lolentz force and is given by [5, 6]

$$m(dx^2/dt^2) + m\gamma(dx/dt) + m(\omega_0^2 x) + (e/c)vH = -eE(t) \tag{5.22}$$

From a solution of this equation for the propagation parallel to an applied magnetic field H, the rotation for low absorption ($n \gg k$) is given by

$$\theta = (\omega d/2c)(n^- - n^+)$$
$$= (\omega d/2c)[(n(\omega_0 + \omega_L) - n(\omega_0 - \omega_L)] \tag{5.23}$$

where ω_L is the 'Larmor frequency'. This difference of the refractive indices for right- and left-circularly polarized light is brought about by the Zeeman splitting of the degenerate energy levels by the magnetic field. The Zeeman splitting causes a slight shift in absorption peaks for right-hand and left-hand circularly polarized light. This shift yields slightly different indices of refraction for the two orthogonal circular polarizations even at wavelengths for which the material is transparent. Figure 5.1 summarizes the relationship between the deviation from the resonance frequency and the Zeeman splitting of energy levels. By selection rules, the change in the M_L component at the transition is $\Delta M_L = 0, +1$ or -1. When the magnetic field is present, the frequency of the emission or absorption line (ω_0 in zero magnetic field) becomes, therefore,

$$\Delta M_L = 0, \ \omega = \omega_0$$
$$\Delta M_L = -1, \ \omega = \omega_0 + \omega_L = \omega_0 - (\mu_B H/\hbar)$$
$$\Delta M_L = +1, \ \omega = \omega_0 - \omega_L = \omega_0 + (\mu_B H/\hbar)$$

corresponding to the left-hand ($\Delta M_L = -1$) and right-hand ($\Delta M_L = +1$) circularly polarized light (perpendicular to the direction of magnetic field), respectively [7].

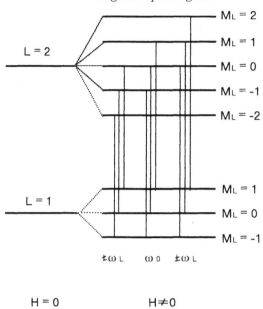

Fig. 5.1.   Relationship between the deviation from the resonance frequency and the Zeeman splitting of energy levels and the sign of the rotation of circularly polarized light.

5.2.2 Verdet constant [8, 9]

Generally the origin of rotation is considered to be of two types depending upon whether the ion or ions in question have a filled or unfilled electronic configuration. In the former case the rotation that is the contribution of the Zeeman effect is termed diamagnetic (positive). Most normal network former and modifier ions in glass would give rise to diamagnetic rotation. In the latter case, the Verdet constant that arises from the perturbation of the amplitude elements of the electric moment by the magnetic field is termed paramagnetic (negative) and the rare-earth ions are examples that are frequently incorporated in glass.

Since the Faraday rotation is brought about by the difference of the refractive indices due to the splitting of energy levels, the expression for the Verdet constant can be written in the form of the index dispersion and more fundamentally on the position and strength of the transition involved. The diamagnetic terms are given by [8, 9]

$$V = (4\pi N \nu^2) \sum_n \{A_n/(\nu^2 - \nu_n^2)^2\} \qquad (5.24)$$

where N is the charge carriers per unit volume, A_n is relative to the oscillator strength of transition, ν_n is the transition frequencies and ν is the frequency of the incident light. This equation can also be written in the form [10] that Becquerel derived on the basis of classical electromagnetic theory,

$$V = (e\nu/2mc^2)(\partial n/\partial \nu) \tag{5.25}$$

where n is the refractive index of the medium and $(\partial n/\partial \nu)$ is the refractive index dispersion. The index dispersion, i.e. the parameter ν_n and A_n would mainly characterize the Verdet constant in this case.

On the other hand, the Verdet constant for the paramagnetic term can be written as [9, 11, 12]

$$V = (K/T)(Nn_{\text{eff}}^2/g)\sum_n\{C_n/(\nu^2 - \nu_n^2)\} \tag{5.26}$$

$$K = 4\pi^2\mu_B\nu^2/3chk \tag{5.27}$$

$$n_{\text{eff}} = g[J(J+1)]^{1/2} \tag{5.28}$$

where N is the number of paramagnetic ions per unit volume, g is the Lande splitting factor, μ_B is the Bohr magneton number, C_n is related to the transition probabilities, ν_n is the transition frequencies, h is the Planck's constant, J is the total angular momentum quantum number, k is the Boltzmann's constant, T is the absolute temperature and the other symbols have the same meaning as above. Therefore, the parameters N, n_{eff}, C_n, and ν_n would characterize the Verdet constant in this case.

5.3 Faraday effect in glasses

5.3.1 Diamagnetic glasses

Most normal network former and network modifier ions in the optical glass, e.g. Si^{4+}, Na^+, Ca^{2+}, Pb^{2+}, and Ba^{2+}, have a filled electron configuration. They would give rise to diamagnetic rotation. The Faraday rotation of optical glasses has been extensively investigated in the visible spectrum and the infrared spectral region.

According to Eq. (5.24), a more convenient procedure for expressing the Verdet constant and dispersion is to plot the Verdet constant against Abbe number ν_d. A correlation of this type has been shown to hold in the visible region for a large number of optical glasses [13]. Figure 5.2 shows the linear relation of Verdet constant as a function of the index difference ($n_F - n_C$) for a number of optical glasses, commercial Faraday glasses and acousto-optic glasses.

Of the optical glasses, SF glasses have a large Verdet constant and are particularly useful for the visible-near infrared spectral region since they have a

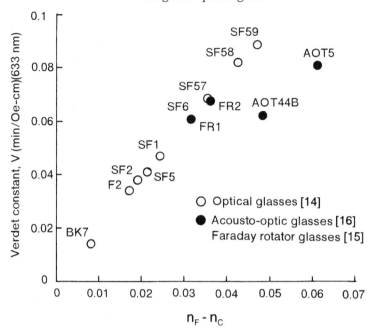

Fig. 5.2.   Relation between Verdet constant and index difference ($n_F - n_C$) of various diamagnetic glasses. (Data from [14–16].)

large dispersion, which is primarily due to the Pb^{2+} ion. Oxide glasses that contain ions with similar electron structure, i.e. a large and easily polarizable outer electron shell, might also have large Verdet constants [9]. These ions are: Tl^+, Sn^{2+}, Pb^{2+}, Sb^{3+}, Bi^{3+} and Te^{4+}. For instance, an $82Tl_2O : 18SiO_2$ (wt%) glass showed a Verdet constant of roughly twice that of SF optical glass. At the other extreme, chalcogenide glasses have very high dispersion and corresponding large Verdet constants. The Verdet constants of various glasses are summarized in Table 5.1.

   In some applications the temperature dependence of the Faraday effect of the material may also be significant. The temperature dependence of the Verdet constant of diamagnetic glass is very small except for the contribution from the thermal expansion of the glass. The Verdet constant of diamagnetic glass FR2 was in fact not affected by the temperature change from 12 to 90 °C [21]. Measured values of $(dV/dT)(1/V)$ for optical glass SF57 [22] and As_2S_3 fiber [19] were approximately 10^{-4}/K and 10^{-2}/K respectively.

5.3.2 Paramagnetic glasses

Paramagnetic glasses that contain rare-earth ions in high concentrations have been extensively investigated [9, 23–30] because some of these glasses have exceptionally large Verdet constants and also have large regions of transparency in the spectral region of interest. Of course, the magnitudes of Verdet constants vary depending upon the choice of the paramagnetic ion.

Converting frequency to wavelength with $\lambda = c/\nu$ and $\lambda_n = c/\nu_n$, and assuming that all the resonant wavelengths λ_n are grouped quite close to the effective transition wavelength λ_t, Eq. (5.26) can be written in the simplified form [23–25]

$$V(\lambda) = (A/T)(Nn_{\mathrm{eff}}^2/g)\{C_t/\{1 - (\lambda^2/\lambda_t^2)\}\} \qquad (5.29)$$

where A is a parameter independent of wavelength. The wavelength λ_t represents a weight average of the actual transition wavelengths which contains the effects of all contributing transitions. This equation means that the Verdet constant is a function of various parameters: the effective number of Bohr magnetons n_{eff}, the effective transition wavelength, the effective transition probability C_t, all of which depend on the choice of rare earth ions and rarely on the host glass matrix, and in addition, on the temperature T and concentration N.

Attempts have been made to correlate the Verdet constants of the rare-earth phosphate and borate glasses with the parameter n_{eff}^2/g arising from Eq. (5.29) [9, 23, 24]. Figure 5.3 shows a plot of the Verdet constant as a function of

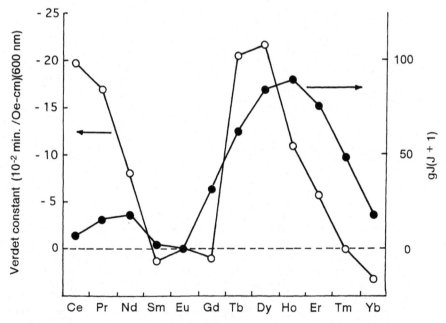

Fig. 5.3. Comparison of V ($\lambda = 600\,\mathrm{nm}$) with n_{eff}^2/g for various trivalent rare-earth ions. [Reprinted from S. B. Berger, C. B. Rubinstein, C. R. Kurkjian and A. W. Treptow, *Phys. Rev.* **133** (1964) A723, copyright (1964) with permission from The American Physical Society.]

Table 5.1. *Verdet constant of various diamagnetic glasses*

Glass	Verdet constant (min/Oe-cm)					Refs.
	(500 nm)	(546 nm)	(633 nm)	(700 nm)	(1.06 μm)	
Optical glass						
SF59		0.128	0.089		0.028	14
SF58		0.118	0.082		0.026	14
SF57		0.099	0.069		0.023	14
SF6		0.087	0.061		0.021	14
SF1		0.068	0.047		0.017	14
SF5		0.058	0.041		0.014	14
SF2		0.053	0.038		0.013	14
F2		0.047	0.034		0.012	14
BK7		0.020	0.014		0.006	14
SFS6	0.166		0.103[b]	0.071	0.032[c]	17
Acousto-optic (TeO$_2$) glass						
AOT44B	0.144[a]		0.081			16
AOT5	0.115[a]		0.0615			16
Faraday rotator glass						
FR1			0.061		0.021	15
FR2			0.068		0.028	15

Glass					Ref
High index glass					
EDF			0.048^d		18
SFS6			0.070^d		18
Corning 8363 (unannealed)			0.074^d		18
Corning 8363 (annealed)			0.072^d		18
$75TeO_2-25Sb_2O_3$			$0.076*$	$0.032*$	9
$20TeO_2-80PbO$			$0.128*$	$0.048*$	9
$84TeO_2-16BaO$			$0.056*$	$0.028*$	9
$75Sb_2O_3-Cs_2O-5Al_2O_3$			$0.074*$	$0.024*$	9
$82Tl_2O-18SiO_2$			$0.100*$	$0.042*$	9
$95Bi_2O_3-5B_2O_3$			$0.085*$	$0.032*$	9
$95PbO-5B_2O_3$			$0.093*$	$0.032*$	9
$95Tl_2O-5B_2O_3$			$0.091*$	$0.040*$	9
Chalcogenide glass					
As_2S_3		0.298^b		0.081^c	17
As_2Se_3				0.110^c	17
As_2S_3				0.0162^f	19
$As_{40}S_{60}$	$0.22**$	$0.28**$	$0.21**$		20
$As_{20}S_{80}$		$0.12**$	$0.093**$		20
$As_{40}S_{57}Se_3$		$0.31**$	$0.23**$		20
$Ge_{20}As_{20}S_{60}$		$0.20**$	$0.155**$		20

[a] 497 nm; [b] 600 nm; [c] 1 μm; [d] 693 nm; [f] 3.39 μm; * estimated from data (given in a unit of rad/T-m) in ref [9]. ** estimated from data in ref [20].

atomic number, which may be compared with the quantity n_{eff}^2/g for each of the rare earth (III) ions in the phosphate glasses [23]. The agreement in shape between Verdet constant and n_{eff}^2/g is satisfactory. Among the trivalent ions, Tb^{3+}, Dy^{3+} and Ce^{3+} give rise to a large Verdet constant. Moreover, these ions also have large regions of transparency in the visible and near infrared spectral region.

The Verdet constant is approximately linear with concentration of paramagnetic ions. High rare-earth containing glasses should, therefore, yield a large Verdet constant. However, the concentration of rare-earth ions that can be added to a glass is limited by its glass-forming properties. Many attempts were made subsequently between 1960 and 1970 to increase rare-earth ions in various glass systems such as phosphate [9, 23, 27], borate [9, 24], silicate [9], aluminoborate [25, 26], alumino-silicate [28, 29] and fluorophosphate [30]. Figure 5.4 shows the typical examples of the glass formation region in the

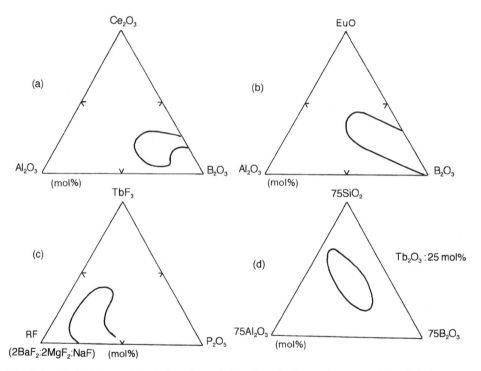

Fig. 5.4. Typical examples of glass formation regions in the systems containing large amounts of rare-earth ions (mol%). (a) Ce_2O_3-aluminoborate (b) EuO-aluminoborate (c) TbF_3-fluorophosphate (d) Tb_2O_3-aluminoborosilicate. [(b) reprinted from M. W. Shafer and J. C. Suits, *J. Am. Ceram. Soc.* **49** (1966) 261, with permission of The American Ceramic Society, Post Office Box 6136, Westerville, Ohio 43086-6136, copyright (1966) by The American Ceramic Society. All rights reserved.]

system containing large amounts of rare-earth ions. Table 5.2 summarizes the Verdet constant of paramagnetic glasses containing various rare-earth ions. Using the data in this table, the Verdet constant as a function of rare-earth concentration is plotted in Fig. 5.5.

Since the Verdet constants of diamagnetic ions and paramagnetic ions have opposite signs, the glass-forming system itself for paramagnetic glasses should be those having low dispersion, i.e. small diamagnetic Verdet constant, and be capable of incorporating large quantities of rare-earth ions.

The other parameters such as transition probability C_t and effective transition wavelength λ_t also depend upon rare-earth ions and the glass composition,

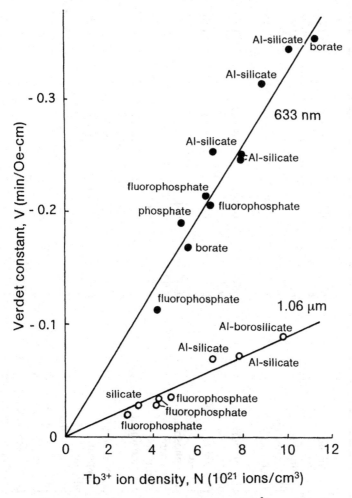

Fig. 5.5.   Verdet constant as a function of Tb^{3+} concentration.

Table 5.2. *Verdet constant of paramagnetic glasses containing rare-earth ions*

RE ion / Host glass	N ($10^{21}/\mathrm{cm}^3$)	Verdet constant (min/Oe-cm)					Refs.
		(400 nm)	(500 nm)	(633 nm)	(700 nm)	(1.06 μm)	
Ce^{3+}							
phosphate	6	−0.672[a]	−0.326	−0.173[b]	−0.132		23
aluminoborate	8.33			−0.22			26
aluminoborate	6.7			−0.19			26
FR4(phosphate)						−0.031	31, 32
Silicophosphate	4.8	−0.58		−0.137		−0.031	33
Pr^{3+}							
lanthan borate	5.0	−0.380[a]	−0.220	−0.128[b]			24
borate	9.2				−0.203	−0.060	9
phosphate	5.3	−0.447[a]	−0.261	−0.150[b]	−0.123		23
silicate	3.79				−0.072	−0.027	9
silicophosphate	4.85	−0.43					33
phosphate	5.92	−0.49					33
aluminoborate	6.64	−0.61					33
aluminosilicate			−0.43[*1]				28
Tb^{3+}							
lanthan borate	5.5	−0.512[a]	−0.288	−0.167[b]	−0.150		24
phosphate	5.4	−0.560[a]	−0.323	−0.190[b]	−0.072		23
silicate	3.37						9
fluorophosphate	6.25			−0.213[*2]		−0.027	30
FR7(fluorophosphate)	4.4			−0.12		−0.033	34
FR7(fluorophosphate)	4.75					−0.035	35
GS309(fluorophosphate)	2.83					−0.019	35
GS320(fluorophosphate)	4.11					−0.027	35
fluorophosphate	6.52			−0.205[*3*5]			36
FR5(aluminosilicate)				−0.242		−0.069	32
FR5(aluminosilicate)						−0.071	31

							Ref.
FR5(aluminosilicate)	7.8			−0.250		−0.071	34, 35
FR5(aluminosilicate)				−0.25			37
FR5(aluminosilicate)			−0.437[f]	−0.245			16
EY 1(silicate)				−0.144		−0.041	32
EY 1(silicate)						−0.0406	38
aluminosilicate	8.8			−0.313			39
aluminosilicate	6.6			−0.253[*5]		−0.069[*5]	40
borosilicate	3.8	−0.26					33
barium-borate	6.5	−0.55					33
aluminosilicate	10[*4]			−0.344[*4][*5]			41
ZrF$_4$-fluoride				−0.028[*5]			42
borate	11.2		−0.632	−0.353			43
MOG101(Na-borosilicate)						−0.03	44
MOG04(Al-borosilicate)						−0.073	44
MOG10(Al-borosilicate)	9.8					−0.088	44
aluminosilicate		−0.5[*1]		−0.216			28
Dy^{3+}							
borate	5.8	−0.436[a]	−0.273	−0.159[b]			24
phosphate	6.2	−0.540[a]	−0.331	−0.197[b]	−0.159		23
silicate	3.46				−0.067	−0.032	9
fluorophosphate	4.6			−0.153[*3][*5]			36
BaF$_2$-MnF$_2$-ThF$_4$-fluoride				−0.0136			45
silicophosphate	4.52	−0.42					33
aluminoborate	8.6	−0.93					33
aluminosilicate	8[*4]			−0.241[*4][*5]			41
aluminosilicate		−0.40[*1]					28
Eu^{2+}							
aluminoborate	4.1	−1.18[e]	−0.298	−0.113[d]	−0.091		25
aluminoborate		−1.26		−0.366[d]	−0.268		25
sodiumborate	4.2	−0.401					46
sodiumborate	3.1	−0.241					46
borate		−0.798					46

a 405 nm; b 635 nm; c 620 nm; d 650 nm; e 406 nm; f 502 nm; *1 estimated from data in ref. [28]; *2 estimated value, *3 estimated from data in ref. [36]; *4 estimated from data in ref. [41]; *5 The original data in ref. are given in a unit of rad/m-T.

Table 5.3. *Values of N, C_t, $gJ(J+1)$ and λ_t for typical rare earth ions in various host glasses*

RE ion	Host glass	N (10^{21}/cm³)	V (Min/Oe-cm)	C_t (10^{38} erg-cm³)	$gJ(J+1)$	λ_t (nm)	Ref.
Ce³⁺	phosphate	6.0	−0.33 (500 nm)	28	7.5	289	25, 47
	FR4, phosphate	6.0	−0.326 (500 nm)			289	23
	silicophosphate		−0.09 (633 nm)			280	31
						280	33
	ZrF₄−fluoride	3.8	−0.43 (400 nm)			250	42
	sulfide	1.623				416	48
Pr³⁺	phosphate	5.3	−0.26 (500 nm)	28	16	210	25, 47
	silicophosphate	5.3	−0.261 (500 nm)			210	23
		3.76				220	33
	lanthan−borate	5.0	−0.33 (400 nm)			210	24
	ZrF₄−fluoride		−0.220 (500 nm)			190	42
	sulfide	1.627				328	48
Tb³⁺	phosphate	5.4	−0.32 (500 nm)	7.8	63	215	25, 47
	lanthan−borate	5.4	−0.323 (500 nm)			215	23
		5.5	−0.288 (500 nm)			225	24
	fluorophosphate	6.25	−0.205 (633 nm)*1			216.8	36
	borosilicate	3.8	−0.26 (400 nm)			250	33
	FR5, alumino−silicate	7.8				219	31
	alumino−silicate	6.6	−0.254 (633 nm)*2			385	40
	borate	11.2	−0.632 (500 nm)	7.33		255	43

ZrF$_4$–fluoride	1.637			191	42
sulfide	1.1			277	48
AlF$_3$–fluoride	1.067			218	49
AlF$_3$–ZrF$_4$–fluoride	0.806			219	49
fluorophosphate	1.951			234	49
borate				220	49
Dy^{3+} phosphate	6.2	8.5	−0.33 (500 nm)	175	25, 47
lanthan–borate	6.2		−0.331 (500 nm)	175	23
	5.8		−0.273 (500 nm)	180	24
fluorophosphate	4.6		−0.153 (633 nm)[*1]	173	36
silicophosphate	3.55		−0.37 (400 nm)	190	33
ZrF$_4$–fluoride				165	42
sulfide				253	48
Eu^{2+} aluminoborate	1.635	31.5	−0.298 (500 nm)	384	25, 47
sodiumborate	4.1	3.0	−0.241 (500 nm)	386	46
sodiumborate	3.1	3.2	−0.486 (500 nm)	417	46
sulfide	3.4	3.8		478	49
AlF$_3$–fluoride	1.645			356	50
sodium borate	1.12			428	50
sulfide	2.05			505	50
	0.89				

[*1] estimated from data (given in a unit of rad/m-T) in ref. [36]; [*2] The original data are given in a unit of rad/m-T.

i.e. the ligand field around the rare-earth ions. From the relationship of Eq. (5.29), a plot of $(1/V)$ versus λ^2 should yield a straight line for the paramagnetic Verdet constant. Figure 5.6 shows the relationship between V^{-1} and λ^2 for typical glasses. The values of C_t and λ_t can be obtained from its slope and the intercept of the horizontal axis, respectively. The values of N, C_t, $gJ(J + 1)$ and λ_t for typical rare-earth ions in various host glasses are listed in Table 5.3.

The effective transition wavelength, λ_t has been identified as arising from the permitted electron-dipole transition of the 4f electron to the 5d shell [9, 23, 24]. Of the rare-earth ions, Ce^{3+} and Eu^{2+} have a relatively large Verdet constant due to their large effective transition wavelength [9, 23–25]. However, Ce^{3+} and Eu^{2+} have the stable 4+ and 3+ state of valency, respectively. A reducing melting condition is required [25, 26] to keep the state of lower valency even in the phosphate glasses. Cerium oxalate $(Ce_2(C_2O_4)_3)$ rather than cerium oxide CeO_2 is preferably used as the raw material. In contrast, the large Verdet constant of Tb^{3+} is primarily due to a large J value.

The effective transition wavelength of, for instance, a Tb^{3+} ion in the fluorides [42, 49] is significantly smaller than those observed in comparable Tb^{3+}-oxide glasses. This difference may be attributed to the higher energy of

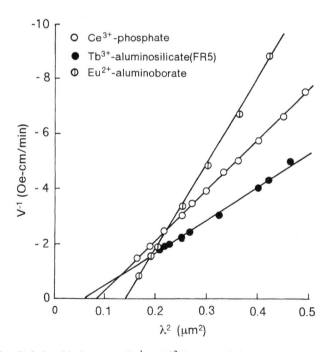

Fig. 5.6.   Relationship between V^{-1} and λ^2 for typical glasses. (Data from [15, 23, 25].)

Table 5.4. *Temperature dependence of Verdet constant of paramagnetic glasses*

Rare-earth ion	Concentration (10^{20} ions/cm^3)	(V/T) ($\lambda = 632.8$ nm) (min-K/Oe-cm)	Ref.
(Silicate)			
Pr	38.0	27	12
Nd	37.2	10	12
Tb	39.0	27	12
Dy	34.6	35	12
Ho	35.0	15.5	12
(Borate)			
Pr	92.0	83	12
Nd	94.2	36	12
(Borosilicate)			
Tb	78	75.3	37
(Aluminosilicate)			
Tb	88	90	39

the 5d orbital and a small splitting of 5d levels of Tb^{3+} ions in fluoride glasses [49]. The value of λ_t for Tb^{3+} ion is larger in borate glasses [43] than in the other oxide glasses. The value of λ_t is much larger in sulfide glasses but the effect of the diamagnetism of the glass system on the Verdet constant is not negligible in this glass system [48].

Equations (5.26) or (5.29) show that the paramagnetic Verdet constant also increases with the reciprocal of absolute temperature in K. Table 5.4 summarizes the variation of V with $1/T$ for the rare-earth ions reported in the literature [12, 37, 39]. At room temperature, the peak angle of rotation increases linearly with magnetic field [38] up to 60 kOe according to Eq. (5.20). At low temperature, however, the magnetic field dependence of the Faraday rotation exhibits an unusual nonmonotonic behavior and has a saturation value [37]. This is expected from the theoretical treatment since the Faraday rotation is linearly dependent on both the magnetization and the applied magnetic field.

5.3.3 Applications

5.3.3.1 Optical isolator (paramagnetic glasses)

An optical isolator is a unidirectional light gate, which can prevent back reflection of light into lasers or optical amplifiers and reduce signal instability and noise [18]. In the high-power laser fusion experiments, for instance, laser

materials or optical components must be protected from back-scattered light
[32, 51]. In optical communication systems, optical isolators are required to
stabilize the oscillation of laser diodes [52, 53].

A schematic diagram of the optical isolator is shown in Fig. 5.7. A Faraday
rotation material is placed between two polarizers (input and output) oriented
at 45° with each other and a magnetic field is applied to rotate the plane of
polarization of light traversing the glass by 45°. Forward-going light is linearly
polarized at the input polarizer, rotated 45° on passage through the glass and
transmitted by the output polarizer. Back-reflected light is linearly polarized at
the output polarizer; however, it undergoes an additional 45° rotation during its
passage through the glass and is rejected by the input polarizer.

Compact size and production of a large Faraday rotation angle with little
power consumption of the magnetic system are the most desirable features of
Faraday isolators. The Verdet constant of the material should be as large as
possible to reduce L and the absorption. A good Faraday rotator material also
includes low absorption at the laser wavelength, low birefringence and high
homogeneity. Miniaturized isolators have been proposed using a new configuration for a fiber isolator consisting of micropolarizers and Tb^{3+}-doped fiber
rotators [54].

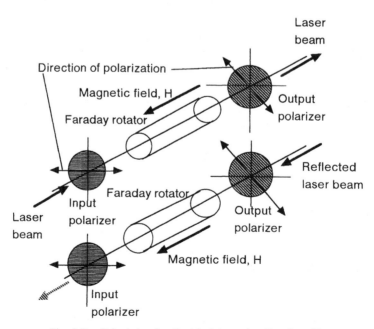

Fig. 5.7.   Principle of optical isolator using Faraday effect.

5.3.3.2 Current sensor (diamagnetic glasses)

Current sensors utilizing the Faraday effect have attractive features such as small size, immunity to noise, a wide dynamic range and high speed response [55, 56]. If a Faraday sensor element in the fiber form is possible, further simple and compact current measurement systems can be developed [57–59].

A current sensor consists of a glass fiber coil turning around an electric power cable. Linearly polarized laser light emitted from a polarization-maintaining fiber is inserted into a polarizer whose azimuth is coincident with the axis of the polarized light. After passing through a half waveplate, the light from the polarizer is focused and launched into the glass fiber coil (for instance 2.5 turns, diameter 8 cm). The light is given a rotation of the polarization plane due to the Faraday effect generated by the current passing through the cable, and emitted from the fiber. The light is divided into two orthogonally polarized beams by an analyzer made of polarization beam splitter, and the rotation of the polarization angle is converted into intensities of light.

These applications require temperature insensitivity over a broad temperature range. Diamagnetic glasses in bulk and fiber form are commonly chosen for the sensing element with high stability, because their temperature dependence is much less than that of paramagnetic materials. Another significant problem arising in fiber-sensing applications is the photoelastic effect induced by bending [57, 60], which usually disturbs the polarization of light in the sensing fiber, i.e. any Faraday rotation. To solve this problem, a single mode fiber is made from special flint glass with no photoelastic effect [59]. By selecting flint glass compositions, its photoelastic constant can be minimized in principle [61, 62].

5 References

1 A. J. Dekker: *Solid State Physics* Maruzen Asian edn, Chap. 18 (Prentice Hall Inc. & Marugen Co. Ltd., 1957) p. 446.
2 C. Kittel: Chap. 14 in *Introduction to Solid State Physics*, sixth edn, (John Wiley & Son, Inc., New York, 1986) p. 395.
3 M. Faraday: *Trans. Roy. Soc.* **120 II** (1830) 1.
4 M. Faraday: *Phil. Trans. Roy. Soc.* **1** (1846) 104.
5 J. G. Mavroides: Chap. 7 in *Optical Properties of Solids*, ed. F. Abeles, (North-Holland Publishing Co., Amsterdam, 1972) p. 351.
6 T. S. Moss, G. J. Burrell and B. Ellis: Chap. 4 in *Semiconductors Opto-Electronics*, (Butterworth & Co. Ltd., London, 1973) p. 95.
7 F. K. Richtmyer, E. H. Kennard and T. Lauritsen: Chap. 3 and 7 in *Introduction to Modern Physics*, fifth edn. (McGraw-Hill Book Company, Inc., New York, and Kogakusha Company Ltd., Tokyo, 1955) p. 77 and 225.
8 R. Serber: *Phys. Rev.* **41** (1932) 489.

9   N. F. Borrelli: *J. Chem. Phys.* **41** (1964) 3289.

10 H. Becquerel: *Compt. Rend.* **125** (1897) 679.

11 J. H. Van Vleck and M. H. Hebb: *Phys. Rev.* **46** (1934) 17.

12 N. F. Bollelli: *Paper of VII International Congress on Glass* (Bruxelles) (1965)
    (Gordon & Research, New York, 1966) paper No 43.

13 H. Cole: *J. Soc. Glass Technol.* **34** (1950) 220.

14 *Schott Optical Glass Catalogue.*

15 *Hoya Catalog* No. 8510–2R.

16 A. B. Villaverde and E. C. C. Vasconcellos: *Appl. Opt.* **21** (1982) 1347.

17 C. C. Robinson: *Appl. Opt.* **3** (1964) 1163.

18 L. J. Aplet and J. W. Carson: *Appl. Opt.* **3** (1964) 544.

19 H. Sato, M. Kawase and M. Saito: *Appl. Opt.* **24** (1985) 2300.

20 J. Qui, H. Kanbara, H. Nasu and K. Hirao: *J. Ceram. Soc. Jpn.* **106** (1998) 228.

21 K. Shiraishi, K. Nishino and S. Kawakami: *Appl. Opt.* **24** (1985) 1896.

22 P. A. Williams, A. H. Rose, G. W. Day, T. E. Milner and M. N. Deeter: *Appl. Opt.*
    30 (1991) 1176.

23 S. B. Berger, C. B. Rubinstein, C. R. Kurkjian and A. W. Treptow: *Phys. Rev.* **133**
    (1964) A723.

24 C. B. Rubinstein, S. B. Berger, L. G. Van Uitert and W. A. Bonner: *J. Appl. Phys.*
    35 (1964) 2338.

25 M. W. Shafer and J. C. Suits: *J. Am. Ceram. Soc.* **49** (1966) 261.

26 Y. Asahara and T. Izumitani: *Proc. 1968 Annual Meeting, Ceramic Assoc. of Jpn.*
    (1968) A10.

27 T. Izumitani and Y. Asahara: *Jap. Pat.* 14402 (1973).

28 C. C. Robinson and R. E. Graf: *Appl. Opt.* **3** (1964) 1190.

29 H. Tajima, Y. Asahara and T. Izumitani: US Pat. 3,971,723 (1976).

30 Y. Asahara and T. Izumitani: *Proc. Electro-Optics/Laser 78*, Boston (1978) p. 16.

31 M. J. Weber: *Laser Program Annual Report – 1976* (Lawrence Livermore
    Laboratory, 1977) p. 2-274.

32 G. Leppelmeir, W. Simmons, R. Boyd, C. Fetterman and J. E. Lane: *Laser Fusion
    Program Annual Report – 1974* (Lawrence Livermore Laboratory, 1975) p.
    160.

33 G. T. Petrovskii, I. S. Edelman, T. V. Zarubina, A. V. Malakhovskii, V. N. Zabluda
    and M. Yu Ivanov: *J. Non-Cryst. Solids* **130** (1991) 35.

34 S. E. Stokowski, M. J. Weber, R. A. Saroyan and G. Linford: *Laser Program
    Annual Report –1977*, Vol. 1 (Lawrence Livermore Laboratory, 1978) p.
    2-165.

35 M. J. Weber: *Laser Program Annual Report – 1978*, Vol. 2 (Lawrence Livermore
    Laboratory, 1979) p. 7-76.

36 V. Letellier, A. Seignac, A. Le Floch and M. Matecki: *J. Non-Cryst. Solids* **111**
    (1989) 55.

37 J. A. Davis and R. M. Bunch: *Appl. Opt.* **23** (1984) 633.

38 J. F. Holzrichter: *NRL Memorandum Report 2510* (Naval Research Laboratory,
    Washington DC, 1972).

39 M. Daybell, W. C. Overton, Jr., and H. L. Laquer: *Appl. Phys. Lett.* **11** (1967) 79.

40 J. Ballato and E. Snitzer: *Appl. Opt.* **34** (1995) 6848.

41 J. T. Kohli and J. E. Shelby: *Phys. Chem. Glass.* **32** (1991) 109.

42 D. R. MacFarlane, C. R. Bradbury, P. J. Newman and J. Javorniczky: *J. Non-Cryst.
    Solids* **213 & 214** (1997) 199.

43 K. Tanaka, K. Hirao and N. Soga: *Jpn. J. Appl. Phys.* **34** (1995) 4825.

44 A. N. Malshakov, G. A. Pasmanik and A. K. Potemkin: *Appl. Opt.* **36** (1997) 6403.
45 O. H. El-Bayoumi and M. G. Drexhage: *J. Non-Cryst. Solids* **56** (1983) 429.
46 K. Tanaka, K. Fujita, N. Soga, J. Qiu and K. Hirao: *J. Appl. Phys.* **82** (1997) 840.
47 J. C. Suits, B. E. Argyle and M. J. Freiser: *J. Appl. Phys.* **37** (1966) 1391.
48 J. Qiu, J. B. Qiu, H. Higuchi, Y. Kawamoto and K. Hirao: *J. Appl. Phys.* **80** (1996) 5297.
49 J. Qiu, K. Tanaka, N. Sugimoto and K. Hirao: *J. Non-Cryst. Solids* **213 & 214** (1997) 193.
50 J. Qiu and K. Hirao: *J. Ceram. Soc. Jpn.* **106** (1998) 290.
51 P. J. Brannon, F. R. Franklin, G. C. Hauser, J. W. Lavasek and E. D. Jones: *Appl. Opt.* **13** (1974) 1555.
52 H. Kuwahara, Y. Onoda, M. Sasaki and M. Shirasaki: *Opt. Commun.* **40** (1981) 99.
53 K. P. Birch: *Opt. Commun.* **43** (1982) 79.
54 K. Shiraishi, S. Sugaya and S. Kawakami: *Appl. Opt.* **23** (1984) 1103.
55 K. Kyuma, S. Tai and M. Nunoshita: *Optics and Lasers in Engineering,* **3** (1982) 155.
56 R. P. Tatam, M. Berwick, J. D. C. Jones and D. A. Jackson: *Appl. Phys. Lett.* **51** (1987) 864.
57 A. M. Smith: *Appl. Opt.* **17** (1978) 52.
58 A. Papp and H. Harms: *Appl. Opt.* **19** (1980) 3729.
59 K. Kurosawa, I. Masuda and T. Yamashita: *Proc. 9th Optical Fiber Sensor Conference, Firenze* (1993) p. 415.
60 D. Tang, A. H. Rose, G. W. Day and S. M. Etzel: *J. Ligthwave Technol.* **9** (1991) 1031.
61 W. A. Weyl and E. C. Marboe: *The Constitution of Glass, II,* (Wiley Interscience, New York, 1964) p. 824.
62 M. Tashiro: *J. Soc. Glass Technol.* (1956) 353T.

Index